EPISTLES OF THE BRETHREN OF PURITY

EPISTLES OF THE BRETHREN OF PURITY

On Geography

An Arabic Critical Edition and English Translation of EPISTLE 4

Edited and Translated by

Ignacio Sánchez and James Montgomery

Foreword by

Nader El-Bizri

OXFORD
UNIVERSITY PRESS
in association with
The Institute of Ismaili Studies

OXFORD
UNIVERSITY PRESS

Great Clarendon Street, Oxford OX2 6DP

Oxford University Press is a department of the University of Oxford.
It furthers the University's objective of excellence in research, scholarship,
and education by publishing worldwide in
Oxford New York
Auckland Cape Town Dar es Salaam Hong Kong Karachi
Kuala Lumpur Madrid Melbourne Mexico City Nairobi
New Delhi Shanghai Taipei Toronto

With offices in
Argentina Aus tria Brazil Chile Czech Republic France Greece
Guatemala Hungary Italy Japan Poland Portugal Singapore
South Korea Switzerland Thailand Turkey Ukraine Vietnam
Oxford is a registered trade mark of Oxford University Press
in the UK and certain other countries

Published in the United States
by Oxford University Press Inc., New York

First published 2014

British Library Cataloguing in Publication Data
Data available

Library of Congress Cataloguing in Publication Data
Data available

ISBN 978-0-19-872822-1

1 3 5 7 9 10 8 6 4 2

Typeset and printed in Lebanon
on acid-free paper by Saqi Books

The Institute of Ismaili Studies

The Institute of Ismaili Studies was established in 1977 with the object of promoting scholarship and learning on Islam, in historical as well as contemporary contexts, and a better understanding of its relationship with other societies and faiths.

The Institute's programmes encourage a perspective which is not confined to the theological and religious heritage of Islam, but seeks to explore the relationship of religious ideas to broader dimensions of society and culture. The programmes thus encourage an interdisciplinary approach to the materials of Islamic history and thought. Particular attention is also given to issues of modernity that arise as Muslims seek to relate their heritage to the contemporary situation.

Within the Islamic tradition, the Institute's programmes seek to promote research in those areas which have, to date, received relatively little attention from scholars. These include the intellectual and literary expressions of Shiʿism in general, and Ismailism in particular.

In the context of Islamic societies, the Institute's programmes are informed by the full range and diversity of cultures in which Islam is practised today, from the Middle East, South and Central Asia, and Africa to the industrialised societies of the West, thus taking into consideration the variety of contexts which shape the ideals, beliefs and practices of the faith.

These objectives are realised through concrete programmes and activities organised and implemented by various departments of the Institute. The Institute also collaborates periodically, on a programme-specific basis, with other institutions of learning in the United Kingdom and abroad.

The Institute's academic publications fall into several distinct and interrelated categories:

1. Occasional papers or essays addressing broad themes of the relationship between religion and society in the historical as well as modern contexts, with special reference to Islam.
2. Monographs exploring specific aspects of Islamic faith and culture, or the contributions of individual Muslim figures or writers.
3. Editions or translations of significant primary or secondary texts.
4. Translations of poetic or literary texts which illustrate the rich heritage of spiritual, devotional and symbolic expressions in Muslim history.
5. Works on Ismaili history and thought, and the relationship of the Ismailis to other traditions, communities and schools of thought in Islam.
6. Proceedings of conferences and seminars sponsored by the Institute.
7. Bibliographical works and catalogues which document manuscripts, printed texts and other source materials.

This book falls into category three listed above.

In facilitating these and other publications, the Institute's sole aim is to encourage original research and analysis of relevant issues. While every effort is made to ensure that the publications are of a high academic standard, there is naturally bound to be a diversity of views, ideas and interpretations. As such, the opinions expressed in these publications must be understood as belonging to their authors alone.

Epistles of the Brethren of Purity

The *Epistles of the Brethren of Purity* is published by Oxford University Press in association with the Institute of Ismaili Studies, London. This bilingual series consists of a multi-authored Arabic critical edition and annotated English translation of the *Rasāʾil Ikhwān al-Ṣafāʾ* (ca. tenth-century Iraq).

Previously published

Nader El-Bizri, ed. *The Ikhwān al-Ṣafāʾ and their 'Rasāʾil': An Introduction* (2008).

Ikhwān al-Ṣafāʾ. *The Case of the Animals versus Man Before the King of the Jinn: An Arabic Critical Edition and English Translation of Epistle 22*, ed. and tr. Lenn E. Goodman and Richard McGregor (2009).

Ikhwān al-Ṣafāʾ. *On Logic: An Arabic Critical Edition and English Translation of Epistles 10–14*, ed. and tr. Carmela Baffioni (2010).

Ikhwān al-Ṣafāʾ. *On Music: An Arabic Critical Edition and English Translation of Epistle 5*, ed. and tr. Owen Wright (2010).

Ikhwān al-Ṣafāʾ. *On Magic I: An Arabic Critical Edition and English Translation of Epistle 52a*, ed. and tr. Godefroid de Callataÿ and Bruno Halflants (2011).

Ikhwān al-Ṣafāʾ. *On Arithmetic and Geometry: An Arabic Critical Edition and English Translation of Epistles 1 & 2*, ed. and tr. Nader El-Bizri (2012).

Ikhwān al-Ṣafāʾ. *On the Natural Sciences: An Arabic Critical Edition and English Translation of Epistles 15–21*, ed. and tr. Carmela Baffioni (2013).

Ignacio Sánchez obtained his PhD from the University of Cambridge in 2011, with a dissertation on al-Jāḥiẓ's treatises on the imamate. He has worked at Cambridge University Library (2010–2012), as part of the team of the Oxford and Cambridge Islamic Manuscripts Catalogue Online Project, and the Digital Library; and at the Institute of Historical Sciences of the Humboldt University in Berlin (2012–2014), on an interdisciplinary project to write an encyclopaedia of mediaeval endowments and charitable foundations. Currently he is a research fellow at Warwick University, where he works on the Oxford/Warwick Ibn Abī Uṣaybiʿa Project.

James E. Montgomery is Sir Thomas Adams Professor of Arabic at the University of Cambridge, a Director of Studies and Fellow of Trinity Hall. He held a Lectureship at Glasgow and Senior Lectureships at Oslo and Leeds before joining Cambridge in 1997. He founded the journal *Middle Eastern Literatures* and its monograph series *Studies in Arabic and Middle Eastern Literatures*. He also contributed to setting up the Third Edition of the *Encyclopaedia of Islam*, and delivered the prestigious Rosenthal Memorial Lecture in Ancient and Near Eastern Civilisations at Yale in 2008. His most recent publications include *Al-Jāḥiẓ: In Praise of Books* (Edinburgh: Edinburgh University Press, 2013).

Contents

Acknowledgements

This book represents the work of Dr Ignacio Sánchez. I began the edition many years ago by collating the majority of the manuscripts but was frustrated in my attempt to complete it by University responsibilities. I am delighted that my former graduate student, Dr Sánchez, has not only taken the work on, but taken it over and produced his own excellent edition and translation. I have enjoyed collaborating with him on the project in the capacity of an unnecessary mentor. That I am listed on the title page of the book is an honour for me, and a supererogatory act by Dr Sánchez for which I am deeply grateful. I am also deeply grateful to the Institute of Ismaili Studies for their patience during the much overdue gestation of this volume.

James E. Montgomery

I would like to express my deepest gratitude to Professor James Montgomery, to whom I am indebted in so many ways for his invaluable guidance and advice. James involved me in this project when he had already done a great amount of work; since then, we have collaborated at all the stages of the production of the book. It has been a real pleasure to work with him on the edition, translation and study of this epistle, and to benefit from his vast knowledge of classical Islamic intellectual history and his insightful comments. The final result owes as much to him as it does to myself, although I alone am responsible for any possible mistakes and shortcomings. I am also grateful to the Institute of Ismaili Studies, and especially to Tara Woolnough, Russell Harris and Isabel Miller for their patience, their careful proofreading and their suggestions, which have greatly improved the quality of the book.

Ignacio Sánchez

Foreword

The Brethren of Purity (Ikhwān al-Ṣafā') were the anonymous members of a fourth-century AH (tenth-century CE) esoteric fraternity of lettered urbanites that was principally based in the southern Iraqi city of Basra, while also having a significant active branch in the capital of the ʿAbbāsid caliphate, Baghdad. This secretive coterie occupied a prominent station in the history of scientific and philosophical ideas in Islam due to the wide intellectual reception and dissemination of diverse manuscripts of their famed philosophically oriented compendium, the *Epistles of the Brethren of Purity* (*Rasāʾil Ikhwān al-Ṣafāʾ*). The exact dating of this corpus, the identity of its authors, and their doctrinal affiliation remain unsettled questions that were hitherto shrouded with mystery. Some situate the historic activities of this brotherhood at the eve of the Fāṭimid conquest of Egypt (ca. 358/969),[1] while others identify the organization with an earlier period that is set chronologically around the founding of the Fāṭimid dynasty in North Africa (ca. 297/909).

The most common account regarding the presumed identity of the Ikhwān is usually related on the authority of the famed *littérateur* Abū Ḥayyān al-Tawḥīdī (ca. 320–414/930–1023), who noted in his *Book of Pleasure and Conviviality* (*Kitāb al-Imtāʿ wa'l-muʾānasa*) that these adepts were obscure 'men of letters': Abū Sulaymān Muḥammad b. Maʿshar al-Bustī (nicknamed al-Maqdisī); the *qāḍī* Abū al-Ḥasan ʿAlī b. Hārūn al-Zanjānī; Abū Aḥmad al-Mihrajānī (also known as Aḥmad al-Nahrajūrī); and Abū al-Ḥasan al-ʿAwfī. Abū Ḥayyān also

1 All dates are Common Era, unless otherwise indicated; where two dates appear (separated by a slash), the first date is Hijri (AH), followed by CE.

claimed that they were the senior companions of a secretarial officer at the Būyid regional chancellery of Basra, known as Zayd b. Rifāʿa, who was reportedly an affiliate of the Brethren's fraternity and a servant of its ministry. Even though this story was reaffirmed by several classical historiographers in Islamic civilization, it is not fully accepted by scholars in terms of its authenticity. Furthermore, some Ismaili missionaries (*duʿāt*) historically attributed the compiling of the *Epistles* to the early Ismaili Imams Aḥmad b. ʿAbd Allāh (al-Taqī [al-Mastūr]) or his father, ʿAbd Allāh (Wafī Aḥmad), while also suggesting that the *Rasāʾil* compendium was secretly disseminated in mosques during the reign of the ʿAbbāsid caliph al-Maʾmūn (r. 198–218/813–833).

Encountering 'veracity in every religion', and grasping knowledge as 'pure nourishment for the soul', the Ikhwān associated soteriological hope and the attainment of happiness with the scrupulous development of rational pursuits and intellectual quests. Besides the filial observance of the teachings of the Qurʾān and Hadith, the Brethren also reverently appealed to the Torah of Judaism and to the Gospels of Christianity. Moreover, they heeded the legacies of the Stoics and of Pythagoras, Hermes Trismegistus, Socrates, Plato, Aristotle, Plotinus, Nicomachus of Gerasa, Euclid, Ptolemy, Galen, Proclus, Porphyry, and Iamblichus.

The Brethren promoted a convivial and earnest 'companionship of virtue'. Their eschatological outlook was articulated by way of an intricate cyclical view of 'sacred' history that is replete with symbolisms and oriented by an uncanny hermeneutic interpretation of the microcosm and macrocosm analogy; namely, believing that the human being is a microcosmos, and that the universe is a 'macroanthropos'. The multiplicity of the voices that were expressed in their *Epistles* reflects a genuine quest for wisdom driven by an impetus that is not reducible to mere eclecticism; indeed their syncretism grounded their aspiration to establish a spiritual refuge that would transcend the sectarian divisions troubling their era.

In general, fifty-two epistles are enumerated as belonging to the *Rasāʾil Ikhwān al-Ṣafāʾ*, and these are divided into the following four parts: Mathematics, Natural Philosophy, Sciences of the Soul and Intellect, and Theology. The first part consists of fourteen epistles, and it deals with 'the mathematical sciences', treating a variety of topics in arithmetic, geometry, astronomy, geography, and music. It also includes five epistles on elementary logic, which consist of the following: the *Isagoge*, the *Categories*, *On Interpretation*, the *Prior Analytics*, and the *Posterior Analytics*. The second part of the corpus groups together seventeen epistles on 'the physical *qua* natural sciences'. It thus treats themes on matter and form, generation and corruption, metallurgy, meteorology, a study of the essence of nature, the classes of plants and animals (the latter being also set as a fable), the composition of the human body and its embryological constitution, a cosmic grasp of the human being as microcosm, and also the investigation of the phonetic and structural properties of languages and their differences. The third part of the compendium comprises ten tracts on 'the psychical and intellective sciences', setting forth the 'opinions of the Pythagoreans and of the Brethren of Purity', and accounting also for the world as a 'macroanthropos'. In this part the Brethren also examined the distinction between the intellect and the intelligible, and they offered explications of the symbolic significance of temporal dimensions, epochal cycles, and the mystical expression of the essence of love, together with an investigation of resurrection, causes and effects, definitions and descriptions, and the various types of motions. The fourth and last part of the *Rasāʾil* deals with 'the *nomic qua* legal and theological sciences' in eleven epistles. These address the differences between the varieties of religious opinions and sects, as well as delineating the 'Pathway to God', the virtues of the Ikhwān's companionship, the characteristics of genuine believers, the nature of the divine *nomos*, the call to God, the actions of spiritualists, of jinn, angels, and recalcitrant demons, the species of politics, the layered ordering of the world, and, finally, the essence of magic and talismanic incantations. Besides the fifty-two tracts that constitute the *Rasāʾil Ikhwān al-Ṣafāʾ*, this compendium was accompanied by a treatise entitled *al-Risāla al-jāmiʿa* (The Comprehensive Epistle), which acted as the *summa summarum* for the whole corpus, and was itself supplemented by a further abridged appendage known as the *Risālat jāmiʿat al-jāmiʿa* (The Condensed Comprehensive Epistle).

In spite of their erudition and resourcefulness, it is doubtful whether the Brethren of Purity can be impartially ranked amongst the authorities of their age in the realms of science and philosophy. Their inquiries into mathematics, logic, and natural sciences were recorded in the *Epistles* in a synoptic and diluted fashion, whilst sporadically being infused with gnostic, symbolic, and occult directives. Nonetheless, their accounts of religiosity, as well as their syncretic approach, together with their praiseworthy efforts to collate the sciences, and to compose a pioneering 'encyclopaedia', all bear signs of commendable originality.

In terms of the epistemic significance of the *Epistles*, and the intellectual calibre of their authors, it must be stated that, despite being supplemented by oral teachings in seminars (*majālis al-ʿilm*), the heuristics embodied in the *Rasāʾil* were not representative of the most decisive achievements in their epoch in the domains of mathematics, natural sciences, or philosophical reasoning. Moreover, the sciences were not treated with the same level of expertise across the *Rasāʾil*. Consequently, this opus ought to be judged by differential criteria as regards the relative merits of each of its epistles. In fairness, there are signs of conceptual inventiveness, primarily regarding doctrinal positions in theology and reflections on their ethical-political import, along with signs of an intellectual sophistication in the meditations on spirituality and revelation.

The *Rasā'il* brims with a wealth of ideas, and constitutes a masterpiece of mediaeval literature that presents a populist yet comprehensive adaptation of scientific knowledge. It is perhaps most informative in terms of investigating the transmission of knowledge in Islam, the 'adaptive assimilation' of antique sciences, and the historical evolution of the elements of the *sociology* of learning through the mediaeval forms of the popularization of the sciences and the systemic attempts to canonize them. By influencing a variety of Islamic schools and doctrines, the Brethren's heritage acted as a significant intellectual prompt and catalyst in the development of the history of ideas in Islam, and, as such, it rightfully deserves the station that has been assigned to it among the distinguished Arabic classics and the high literature of Islamic civilization.

The composition of this text displays impressive lexical versatility, which encompasses the technical idioms of mathematics and logic, the heuristics of natural philosophy, and the diction of religious pronouncements and occult invocations, in addition to poetic verses, didactic parables, and satirical and inspirational fables. Despite the sometimes disproportionate treatment of topics, the occasional hiatus in proofs, irrelevant digressions, or instances of verbosity, such apparent stylistic weaknesses disappear, becoming inconsequential when a complete impression is eventually formed of the architectonic unity of the text as a whole, and of the convergence of its constituent elements as a remarkable *oeuvre des belles lettres*.

Modern academic literature on the *Rasāʾil Ikhwān al-Ṣafāʾ* is reasonably extensive within the field of Islamic studies, and it continues to grow, covering works dating from the nineteenth century up to the present, with numerous scholars attempting to solve the riddles surrounding this compendium. The academic rediscovery of the *Rasāʾil* in modern times emerged through the monumental editorial and translation efforts of the German scholar Friedrich Dieterici between the years 1861 and 1872. Several printed editions of reconstructions of the original Arabic have also been established, starting with the *editio princeps* in Calcutta in 1812, which was reprinted in 1846, then a complete edition in Bombay between 1887 and 1889, followed by the Cairo edition of 1928, and the Beirut editions of 1957, 1983, 1995, and their reprints.[1] Although the scholarly contribution of these Arabic editions of the *Rasāʾil* is laudable, as they valuably sustained research on the topic, they are uncritical in character, and they do not reveal their manuscript sources. Consequently, the current printed editions do not constitute definitive primary-source documentations of this classical text. Given this state of affairs, the Institute of Ismaili Studies (IIS) in London has undertaken the publication (in association with Oxford University Press) of a multi-authored, multi-volume,

1 The principal complete editions of this compendium that are available in print consist of the following: *Kitāb Ikhwān al-Ṣafāʾ wa-Khullān al-Wafāʾ*, ed. Wilāyat Ḥusayn, 4 vols. (Bombay: Maṭbaʿat Nukhbat al-Akhbār, 1305–1306/ ca. 1888); *Rasāʾil Ikhwān al-Ṣafāʾ*, ed. Khayr al-Dīn al-Ziriklī, with two separate introductions by Ṭāha Ḥusayn and Aḥmad Zakī Pasha, 4 vols. (Cairo: al-Maṭbaʿa al-*ʿArabiyya* bi-Miṣr, 1928); *Rasāʾil Ikhwān al-Ṣafāʾ*, ed. with introduction by Buṭrus Bustānī, 4 vols. (Beirut: Dār Ṣādir, 1957); and an additional version of the *Rasāʾil Ikhwān al-Ṣafāʾ*, ed. ʿĀrif Tāmir, 5 vols. (Beirut: Manshūrāt *ʿUwaydāt*, 1995).

Arabic critical edition and annotated English translation of the fifty-two epistles. In preparation for the critical edition, reproductions of nineteen manuscripts were acquired by the IIS, and their particulars can be summarized as follows, with the corresponding Arabic sigla:

(I) Bibliothèque Nationale de France:
MS 2303 (1611 CE): [ر]
MS 2304 (1654 CE): [ز]
MS 6.647–6.648 (AH 695; Yazd): [د]

(II) Bodleian Library, Oxford:
MS Hunt 296 (n.d.): [ج]
MS Laud Or. 255 (n.d.): [ح]
MS Laud Or. 260 (1560 CE): [خ]
MS Marsh 189 (n.d.): [غ]

(III) El Escorial, Spain:
MS Casiri 895/Derenbourg 900 (1535–1536 CE): [س]
MS Casiri 923/Derenbourg 928 (1458 CE): [ش]

(IV) The Istanbul collections (mainly the Süleymaniye and associated libraries):
MS Atif Efendi 1681 (1182 CE): [ع]
MS Esad Efendi 3637 (ca. thirteenth century CE): [ن]
MS Esad Efendi 3638 (ca. 1287 CE): [أ]
MS Feyzullah 2130 (AH 704): [ف]
MS Feyzullah 2131 (AH 704): [ق]
MS Köprülü 870 (ca. fifteenth century CE): [ك]
MS Köprülü 871 (1417 CE): [ل]
MS Köprülü 981 (n.d.): [و]

(V) Königliche Bibliothek zu Berlin:
MS 5038 (AH 600/1203 CE): [ب]

(VI) Tehran, The Mahdavī collection:[1]
MS 7437 (AH 640): [ط]

The reconstruction of the contents of the *Rasāʾil* by way of a critical edition is undertaken through manuscript reproductions that are significantly distanced in time from the original text (or from multiple versions of original textual variations), and these have proved to be traceable to a variety of transmission traditions that cannot be articulated with confidence in terms of a definitive *stemma codicum*, or by seeking to establish an '*Urtext*'. The dexterity of the copyists, their deliberate tampering, or commendable exercise of restraint and relative impartiality, along with their scribal idioms, would have conditioned the drafting of the manuscripts and the accessibility to their varying sources. Such endeavours would also have been influenced by the intellectual impress of the prevalent geopolitical circumstances in which this text was transcribed, in addition to its channels of transmission. By widening the selection of the oldest manuscripts and fragments, based on the period of the copying, the levels of completeness and

1 It is worth noting that these acquisitions by the IIS, which consist of the oldest complete manuscripts, along with significant supplementary fragments of an early dating, were each carefully selected from over one hundred extant manuscripts, which are preserved in thirty-nine libraries and collections, noted in alphabetical order by country, as follows: *Egypt*: Dār al-Kutub, Arab League Library (possibly also in the Arab League offices in Tunis); *France*: Bibliothèque Nationale de France; *Germany*: Königliche Bibliothek zu Berlin, Herzogliche Bibliothek zu Gotha, Eberhard–Karlis–Universität (Tübingen), Leipzig (Bibliotheca Orientalis), München Staatsbibliothek; *Iran*: Muṭahharī Library, Tehran University Central Library, Mahdavī (private) Collection; *Ireland*: Chester Beatty Library; *Italy*: Biblioteca Ambrosiana, Biblioteca Vaticana; *Netherlands*: Bibliotheca Universitatis Leidensis; *Russia*: Institut des Langues Orientales (St Petersburg); *Spain*: Biblioteca del Monasterio San Lorenzo de El Escorial; *Turkey*: Süleymaniye, Aya Sofia, Amia Huseyn, Atif Efendi, Esad Efendi, Millet Library, Garullah, Köprülü, Kütüphane-i ʿUmūmī Defterī, Manisa (Maghnisa), Rashid Efendi (Qaysari), Topkapi Saray, Yeni Cami, Revan Kishk; *United Kingdom*: Bodleian Library, British Library, British Museum, Cambridge University (Oriental Studies Faculty Library), the Institute of Ismaili Studies (including copies from the Hamdani, Zāhid ʿAlī, and Fyzee collections), Mingana Collection (Selly Oak Colleges Library, Birmingham), School of Oriental and African Studies (SOAS); *United States*: New York Public Library, Princeton University Library.

clarity, and the recommendations of past and present scholars who have consulted these collections, a suitably grounded critical edition will be produced, and a more reliable textual reconstruction will offer us improved access to the contents of the *Rasāʾil* beyond what is presently available through the printed editions (i.e., those from Bombay, Cairo, and Beirut). It is ultimately hoped that the collective authorial effort, in establishing the Arabic critical edition of the *Rasāʾil* and the first complete annotated English translation, will eventually render service to the academic community and lay a scholarly foundation for further studies dedicated to the Brethren's corpus and its impact on the history of ideas in Islam and beyond.

The present volume is the seventh of the definitive OUP–IIS *Epistles of the Brethren of Purity* series, which was inaugurated in 2008 with the publication of an introductory volume of studies on the Ikhwān al-Ṣafāʾ and their *Rasāʾil*.[1] This volume comprises the Arabic critical edition and annotated English translation with commentaries of Epistle 4 'on geography' from the first part of the *Rasāʾil Ikhwān al-Ṣafāʾ*,[2] which has been prepared by Dr Ignacio Sánchez and Professor James Montgomery. The Brethren of Purity classified geography as one of the noble sciences (*al-ʿulūm al-sharīfa*) that reveals the inherent order of the universe. Ptolemaic mathematical geography along with Pythagorean onto-theology allowed them to have a perspective on 'the rationality of divine creation', and their scientific endeavour gave expression to their piety and ethical pursuits. In this epistle, they described the earth, its spherical shape, the theories of its 'immobility', and the influence of the stars on it. They also examined the topographical features of the land, the equator, the cardinal points, the inhabited quarters, the boundaries of the climes, and the order of cities.

The *Epistles of the Brethren of Purity* series, and this present volume, would not have been feasible without the scholarly contributions of numerous esteemed colleagues who are participating in this challenging textual endeavour to establish the Arabic critical edition and annotated

1 Nader El-Bizri, ed., *The Ikhwān al-Ṣafāʾ and their Rasāʾil: An Introduction* (Oxford: OUP–IIS, 2008).

2 Previously published books in the OUP–IIS *Epistles of the Brethren of Purity* series are mentioned in the preliminary title pages of this present volume.

English translation with commentaries of the *Rasā'il*. I am deeply thankful to all of them for their furtherance of the realization of this venture, and I am specifically beholden to Dr Ignacio Sánchez and Professor James Montgomery for bringing this volume to publication. I also express my gratitude to the distinguished members of the Editorial and Advisory Boards of the series for their continual academic support, especially to Professors Hermann Landolt, Wilferd Madelung, Ismail K. Poonawala, and Roshdi Rạshed. Profound thanks must go as well to my dear colleagues at the Institute of Ismaili Studies, London, and to its Directors and Governors, past and present, for their generous sponsorship of this scholarly project. I would like to record here my indebtedness to Dr Farhad Daftary for his constant support of this institutional initiative. Most special thanks are due to Dr Isabel Miller, Mr Russell Harris, Ms Tara Woolnough, and Mr Saleh al-Achmar for their editorial care.

Nader El-Bizri
(General Editor, *Epistles of the Brethren of Purity*)
Beirut, July 2014

Introduction

1. Text

1.1. Manuscripts

1.1.1. Copies

The version of the Arabic text presented in this edition takes into account twelve manuscripts, whose dates of composition range from the twelfth to the fifteenth centuries:

Atif Efendi, MS 1681	ع	577 AH/1182 CE
Esad Efendi, MS 3638	أ	ca. 686 AH/1287 CE
Esad Efendi, MS 3637	ن	ca. 7th c. AH/13th c. CE
Köprülü, MS 870	ك	855–886 AH/1451–1481 CE
Köprülü, MS 871	ل	820 AH/1417 CE
Feyzullah, MS 2130	ف	704 AH/1305 CE
Tehran (Mahdavī), MS 7437	ط	640 AH/1243 CE
Bibliothèque nationale, MS 2303	ر	1020 AH/1611 CE
Bibliothèque nationale, MS 2304	ز	1064 AH/1654 CE
Bodleian Library, Oxford, MS Hunt 296	ج	Undated
Bodleian Library, Oxford, MS Laud Or. 260	خ	967 AH/1560 CE
Bodleian Library, Oxford, MS Marsh 189	غ	Undated

The oldest identified manuscript of the *Epistles* publically available in libraries is Atif Efendi 1681 [ع]. This manuscript, which contains the complete text of fifty-one epistles, has been the preferred basis for the other editions published to date in this collection.[1] As far as the epistle 'On Geography' is concerned, however, Atif Efendi 1681 does not represent the best tradition of the text: this manuscript has significant omissions, there are a number of instances of textual error, especially as regards the amalgamation of lines, and, unlike the rest of the manuscripts which use tabular form, Atif Efendi 1681 uses continuous text to lists the cities and their co-ordinates.

Following the editorial recommendation of Professor Hermann Landolt as regards Epistle 42, this edition of Epistle 4 is principally based on Esad Efendi 3638 [أ]. Esad Efendi 3638 [أ] was written approximately a century later than Atif Efendi 1681 [ع], but has preserved the most complete and readable version of the main text of the epistle 'On Geography', and also one of the best versions of the accompanying cartographical tables. In general, all the manuscripts have preserved a relatively good copy of the text, with only minor lacunae in some, but the variants are often significant with regard to the numerical values and, especially, with regard to the composition of the tables. It should be noted that the undated manuscript Esad Efendi 3637 [ن] seems to have preserved text from at least two manuscript traditions, as the marginal notes often seem to be the result of a later collation of the base text with a second manuscript of the work.

1.1.2. Description of the Manuscripts

Atif Efendi, MS 1681 ('ayn, ع)

This is a copy with significant omissions, occasionally supplied in the margin by a different hand (for instance on fol. 66a). The manuscript is written in clear *naskh* using black ink with 25 lines per page, and fully vocalized. Red rubrics mark some *fuṣūl* divisions (fols. 66a and 66b). The text is fully dotted with exception of some toponyms and

1 See, for instance, C. Baffioni, *On Logic. An Arabic Critical Edition and English Translation of Epistles 10–14* (Oxford: OUP–IIS, 2010).

abjad numerals. As regards the images and the tables, and in contrast to the other copies, this manuscript has only one image representing the spherical Earth with the cardinal points and the seven climes (fol. 61b), and only one table with the names and co-ordinates of the major cities of the world (fol. 64a); the lists of the cities and their co-ordinates belonging to each of the seven climes are given in continuous text, not in tables, as in the rest of the manuscripts.

Esad Efendi, MS 3638 (*alif*, أ)

This manuscript contains the complete text of the epistle. The script is clear *naskh* in black ink, unvocalized, with 35 lines per page; red rubrics mark the section headings and the *fuṣūl* divisions, and there are catchwords at the bottom of each verso page. The text is fully dotted and the use of diacritics is consistent except in some toponyms, where the text contains only the *ductus* without diacritical marks, in some numerals, and, especially, in the rendition of the longitudes and latitudes using the *abjad* numerical system, where the diacritical pointing is rarely used. The manuscript contains two graphical representations of the spherical Earth: one with the equator and the hemispheres (fol. 36a), and a second one with the cardinal points and the seven climes (fol. 36a); the manuscript also contains eight tables with cities and their co-ordinates (fols. 37a–38b). The numerical values are often written in red ink, and sometimes have a line above, especially when using the *abjad* system.

Esad Efendi, MS 3637 (nūn, ن)

This manuscript contains the complete text of the epistle. The main text is written in a fairly inelegant *naskh* in black ink, unvocalized, with 29 lines per page. The section headings are written in *thuluth* using red ink; this script is also used, in black ink, for the names of the cities in the only table in the manuscript. There are catchwords at the bottom of each verso page. The text is not always dotted, the lack of diacritics being especially noticeable in the toponyms and the *abjad* numerical system. The manuscript contains: an image of the Earth with the equator, and another one with the representation of the cardinal points marked on it, the hemispheres (fol. 52b) and the seven climes (fol. 53a); a table with the major cities of the seven climes and their co-ordinates written in a different hand (fol. 54b), and blank spaces left for another six tables (fols. 55b–58b, the last having two blank spaces). Numerical values sometimes have a line above, especially when using the *abjad* system. The marginalia, written in *thuluth* by a different hand, consist of variants from a different textual tradition, revealing that this manuscript was collated with a second copy of the epistle belonging to group B.[1]

Köprülü, MS 870 (kāf, ك*)*

A very fine copy of the epistle, with minor omissions, written in elegant *naskh* in black ink, and with 31 lines per page. Section headings and rubrics are written using red and gold ink, and the text box is framed by a gilded border. The text is not always dotted, with the lack of diacritics especially noticeable in the toponyms and the *abjad* numerical system. The manuscript contains two representations of the spherical Earth, the first one showing the equator and the cardinal points and the hemispheres, and the second one indicating the seven climes; both images are set within a blue square with gilded floral decoration, and a further frame in red ink. There are seven tables with the names of the cities and their co-ordinates; the first six have duplicate columns and repeat the names of the cities, although the repeated columns with the co-ordinates are empty. Numerical values in Arabic numerals and *abjad* letters are written in red or gold ink. It is worth noting that the careful calligraphy and the delicate decoration of this manuscript contrast strongly with the inaccuracy to be found in the layout of the tables and their erroneous contents.

Köprülü, MS 871 (lām, ل*)*

A complete copy of the epistle written in elegant *naskh*, with 23 lines per page; section headings and rubrics are written in different ink. The text is fully vocalized and dotted, with the exception of certain toponyms and *abjad* numerals. The manuscript contains two simple representations of the spherical Earth, the first one showing the equator

1 See below p. 9.

(fol. 45b), and the second one the seven climes (fol. 46a); both images are incomplete, lacking any names. There are seven tables with the names of the cities and the co-ordinates (fols. 48a, 49a–50a); the cells for longitude and latitude are empty and the names of the cities are written by a different hand in *thuluth* script. In fol. 50a there is a square space left blank. The manuscript has numerous corrections in the margins. Both Arabic numerals and *abjad* letters are sometimes written with a different ink and may have a line drawn over them.

Feyzullah, MS 2130 (fāʾ, ف)

A complete copy written in good *naskh* using black ink; 33 lines per page. The text is partially vocalized and fully dotted, with the common exception of certain toponyms and *abjad* numerals. Section headings and *fuṣūl* divisions are written in red ink. The manuscript has two simple representations of the spherical Earth, the first one showing the equator (fol. 41a), and the second one the seven climes and the extension of the diameter of the Earth (fol. 41b). There are seven tables with the names of the cities and their co-ordinates (fols. 43b, 44a, 45a); the last three tables are incomplete, lacking the longitudes and latitudes of the cities. Arabic numerals and *abjad* letters usually have a line drawn above them. There are several marginal corrections and additions, but the ink of the marginalia is sometimes effaced, rendering the text illegible.

Tehran, MS 7437 (ṭāʾ, ط)

A complete copy written in a very cursive *naskh*, with 20 lines per page. The text is partially vocalized and not always dotted, especially the *abjad* numerals, which do not bear diacritical marks. The manuscript has two simple representations of the spherical Earth, the first one showing the equator and the hemispheres (fol. 52a), and the second one the seven climes and the cardinal points (fol. 52b). The lists of the cities and their co-ordinates are presented in tabular form, using three (fol. 54b) and six columns (fols. 55a–58a), but without the actual table lines. Arabic numerals sometimes have a line drawn above them, but the *abjad* letters do not. There are several marginal annotations in a different hand supplying omitted text.

Bibliothèque nationale, MS 2303 (rāʾ, ر)
A complete copy of the epistle written in fairly cursive *naskh*, with section headings and rubrics in ink of a different colour, and a double-lined border framing the text box, which has 31 lines per page; there are catchwords at the bottom of each verso page, outside the frame. The text, including the *abjad* numerals, is dotted and partially vocalized. The manuscript has two simple representations of the spherical Earth, the first one showing the equator and the hemispheres (fol. 75b), and the second one the seven climes and the poles (fol. 75a). There are eight tables with the names of the cities and their co-ordinates (fols. 77b–80b); the last three tables are incomplete, lacking the co-ordinates. Arabic numerals are not overlined, but are written in a different ink.

Bibliothèque nationale, MS 2304 (zayn, ز)
This manuscript contains the complete text of the epistle. The main text is written in *naskh* script, partially vocalized and fully dotted, with the exception of certain *abjad* numerals; a triple-lined border frames the text box, which has 31 lines per page; there are catchwords at the bottom of each verso page, outside the frame. The pages have two blank squares left for the two images of the spherical Earth (fols. 31b and 32a), and another eight rectangular blank spaces for the tables (fols. 33a–35b). Numerical values sometimes have a line above. There is marginal annotation, probably written by the same hand, which supplies some omitted text.

Bodleian Library, Oxford, MS Hunt 296 (jīm, ج)
A complete copy written in a very inelegant *naskh*, with a significant omission of two lines in the section describing the cardinal points. The section headings and rubrics are written in a different ink. A single-lined frame surrounds the text box, but the text does not always fit inside these limits, continuing in the margins instead. Each page has 25 lines. The text is partially vocalized and not always dotted, especially the toponyms and the *abjad* numerals, which do not always have diacritical marks. The manuscript has two representations of the spherical Earth, the first one showing the equator, the poles, and the hemispheres (fol. 39a), and the second one the seven climes and the cardinal points (fol. 39b). There

are eight tables with the names of the cities and their co-ordinates (fols. 41a–43a); the two first tables and the two last tables are incomplete, lacking the co-ordinates. Arabic numerals are sometimes overlined and written in a different ink; *abjad* numerals are also written in a different ink, but without a line above them.

Bodleian Library, Oxford, MS Laud Or. 260 (khāʾ, خ)

A copy with significant omissions, written in *naskh* script. The section headings and rubrics are written in a different ink; there are 33 lines per page. A double-lined frame surrounds the text box. There are catchwords at the bottom of each verso page, outside the frame. The text is partially vocalized and fully dotted, with the exception of the *abjad* numerals, which do not always have diacritical marks. The manuscript has two images of the spherical Earth, the first one with the equator, the poles, and the hemispheres (fol. 39a), and the second with the seven climes (fol. 39b). There are eight tables with the names of the cities and their co-ordinates (fols. 40b–43a); the three last tables are incomplete, lacking the co-ordinates. Arabic numerals and *abjad* numerals are written in a different ink, without a line above them.

Bodleian Library, Oxford, MS Marsh 189 (ghayn, غ)

A fine complete copy of the epistle, written in careful *naskh*, with section headings and rubrics in a different ink; 29 lines per page. The manuscript has been paginated — not foliated — by a modern hand. The text is partially vocalized and fully dotted, with the common exception of the *abjad* numerals. The manuscript contains two representations of the spherical Earth, the first one showing the equator, the cardinal points, and the hemispheres (p. 82), and the second one the seven climes (p. 83); it is worth noting that, unlike the images of the Earth drawn in the other copies of the epistle, the Northern Hemisphere occupies here the upper part of the sphere. There are nine tables with the names of the cities and their co-ordinates, though only the first four tables contain the data for longitude and latitude. The first table seems to be a particular addition to this manuscript, as its content, which consists of a list of major Iranian cities, does not correspond to any section of the main text and is not found in any other manuscript. Numerical values in Arabic numerals and *abjad* letters are written in a different ink.

1.1.3. Textual Relationships

The nature of this text considerably limits the possibility of a stemmatic analysis; it is often repetitive, both in terms of its structure and its phraseology, and there are standard patterns of error shared by the majority of the manuscripts that do not shed any light on the textual relations, such as skipping a line, or amalgamating two sentences into one because of an identical word or phrase. These omissions and mistakes are often irrelevant as regards the reconstruction of a *stemma codicum*, since the absent text can often be easily supplied or amended by the scribes; for instance, when the author describes the length of the days in each clime, the increase in length is 30 minutes per clime, so an omission or mistake in the calculation of one of these periods need not necessarily be carried over to further copies, as a careful scribe could easily correct a mistake or supply the omitted text. Similarly, the repetition of the same mistake or mistakes in two manuscripts does not necessarily reveal a relationship, unless a number of identical mistakes indicates otherwise. The accompanying tables with names of cities and co-ordinates are not only incomplete but also dissimilar to such an extent that it is not possible to systematize patterns of variations. These characteristics render it extremely difficult to unravel the relationships between the copies consulted for this edition; it is possible, however, to find a variety of signs of potential textual dependence on the basis of the following criteria:

Paratexts

Title headings and divisions are usually not a reliable characteristic with which to analyse textual relationships. They are extremely variable, and they may not belong to the copied text itself, being the result of later scribal additions. In this case, however, the title of the epistle provides important information as regards the position it occupies in the corpus. We can divide the manuscripts into three groups, according to the numeration given in the closing formulae: those which introduce this epistle as number four [ز, ط,

ك, ل], five [أ, ج, خ, ر, غ, ن], and six [أ, ف]. This order might have been the result of a rearrangement of the epistles that does not necessarily correspond to that followed in the manuscript from which they were copied, but it should at least be taken into account together with other characteristics. In this particular case, presenting this epistle as the sixth is an oddity that could very well suggest a direct or indirect relation between manuscripts أ and ف, although manuscript أ, which introduces this epistle as the sixth, gives the number five in the colophon.

Omissions

There are several lacunae shared by certain manuscripts that may indicate a direct relation between them.

The length of the fourth clime is missing in ج, ر, and ن.

The length of the fifth clime is missing in ج, خ, ر, غ, and ن.

The length of the sixth clime is missing in أ, ج, خ, ر, غ, ف, and ن.

The length of the seventh clime is missing in أ, ج, خ, ر, ط, غ, ف and ن.

In the description of the fourth clime, the sentence:

وأهل هذا الإقليم أعدل الناس طباعاً وأخلاقاً

(*wa-ahl hādha'l-iqlīm aʿdal al-nās ṭabāʿan wa-akhlāqan*) is missing in أ, ج, خ, ر, غ, ف, and ن.

Shared mistakes or variants

The expression أبناء الدنيا (*abnāʾ al-dunyā*) occurs as أعرض الدنيا (*aʿraḍ al-dunyā*) in ز, ط, ك, and ل.

The toponym القاسه, (*al-q-ā-s-y-h*) which other manuscripts render as القادسيّة, (*al-qādisiyya*) occurs in ز, ط, ك, and ل.

The ethnonym التغزغز (*al-tughuzghuz*) occurs as بغرغز (*b-gh-z-gh-z*) in ز, ك, and ل; irrespective of possible variants in the diacritical points, these three manuscripts are the only ones where it occurs without the article.

The length of the fourth clime is 7,800 miles in ز, ط, ك; the first cipher of the number is not visible in ل but its last three are 800.

The length of the fifth clime is 7,400 miles in ز, ط, ك, and ل.
The length of the sixth clime is 8,000 miles in ز, ط, ك; and 800 in ل, although this clearly seems to be a mistake for 8,000.
The length of the seventh clime, 6,600 miles, has only been preserved in ز, ط, ك, and ل; it has been added as a marginal note in ن.

Additions in marginalia

Manuscript ن contains marginal additions which seem to have been copied from a manuscript belonging to a textual tradition different to the one used to copy the main text:
In the description of the fourth clime, the sentence:
وأهل هذا الإقليم أعدل الناس طباعاً وأخلاقاً
(*wa-ahl hādhā al-iqlīm aʿdal al-nās ṭabāʿan wa-akhlāqan*), which is missing in أ, ج, خ, ر, غ, and ف, has been copied in the margin of ن.
The lengths of the fourth, fifth, sixth, and seventh climes are missing in the main text, as with the group أ, ج, خ, ر, غ, ف, but supplied in the margins.

Transpositions

The omissions and mistakes, such as the amalgamation of different toponyms containing the word *bilād* (land) or sections introduced by the verb *yamurr* (passes by), greatly influence the displacement of the terms in the enumerations of lands and cities when describing the climes, and it is difficult to find a clear pattern. Nonetheless, the similarity between manuscripts أ and ف is noticeable in this regard.

Tables

For all their inconsistencies, the tables containing the cities with the co-ordinates can also shed some light on the textual relationships of several of the manuscripts. The textual relationship between the tables, however, does not necessarily imply any correspondence to the main text, since many of the tables were added once the copying of the epistle was finished, and it is possible that they might have been copied from a different source.[1] Two main features can be observed in these tables:

It seems evident that the tables in manuscripts خ and ر belong to the same textual tradition, since the order of the cities is almost the same, the co-ordinates of the cities coincide on most occasions, and the cells in which the co-ordinates are lacking are the same in both manuscripts. Moreover, they share four unique characteristics: the city of Mecca is lacking; they list the city of Maʿd in the second clime; Ḥurmā al-Kubra is rendered respectively as *J-r min al-k-n-d-ā* [خ] and *Ḥ-r min al-k-b-d-ā* [ر]; and one of the cells in the table of the second clime contains the word مقدم (*m-q-d-m*), which is unintelligible in this context.

The tables of manuscripts ك and ل are also very similar and share a number of oddities: they both use the expression *ʿalā al-baḥr mīlan* in the first clime, the term *diyār taliya* instead of *diyār bakr* in the second clime, and the last item in the list of cities of the seventh clime is an unidentified toponym, which does not occur in the rest of the manuscripts: *w-r-m-?-m* [ك], *w-r-?-m* [ل].

Diagrams

There is an oddity in the representation of the Earth that only occurs in three manuscripts. Contrary to the common practice in Islamic cartographical representations, the Northern Hemisphere occupies the upper semicircle of the Earth in the first diagram of manuscripts ف, غ, and ن; and this is also the case in the second diagram of manuscript غ.

If we take into consideration these observations, a comparison of the twelve manuscripts used for this edition reveals that there are at least two different textual traditions that seem to correspond broadly to the two different arrangements of the collection, where the epistle

1 The stages of the copying process are clear in manuscripts ز and ن, where blank spaces were reserved for the diagrams and the tables, which were never added.

'On Geography' bears the numbers five and four respectively, with the noted exceptions of manuscript أ, where it is introduced as the sixth epistle on the title page, but as the fifth in the colophon; and of manuscript ف, where it is introduced as the sixth epistle.

Group A

Subgroup A1: manuscripts أ and ف (where it is entitled the 'sixth epistle').

Subgroup A2: manuscripts ج, خ, ر, ع, غ, ن (in which it is entitled the 'fifth epistle'), of which خ and ر, or at least their tables, seem to be part of the same manuscript tradition.

All the manuscripts with a diagram in which the Northern Hemisphere occupies the upper semicircle, i.e., غ, ف, and ن, belong to this subgroup; this could also be an indication of a separate textual tradition.

Group B

Manuscripts ز, ط, ك, ل (in which it is entitled the 'fourth epistle'), of which ك and ل may be part of a separate subgroup within the same manuscript tradition if we use the tables as the criteria for a decision. A manuscript belonging to group B seems to have been used for the marginal notes in ن.

In addition to the aforementioned characteristics, there are copious divergences between the manuscripts in each group, but the majority of these variants seem to be the result of the mistakes engendered by the repetitive nature of the texts and cannot be treated as systematic errors. In order to study any possible direct relation between the extant copies belonging to both textual traditions, a detailed study of the entire collection of epistles would be required. These conclusions, albeit partial, allow us to propose the following stemmatic classification:[1]

---

1 This stemma broadly corresponds to the one proposed by Owen Wright for the epistle 'On Music'; cf. Wright, *On Music: An Arabic Edition and English Translation of Epistle 5* (Oxford: OUP–IIS, 2010), pp. 2–7.

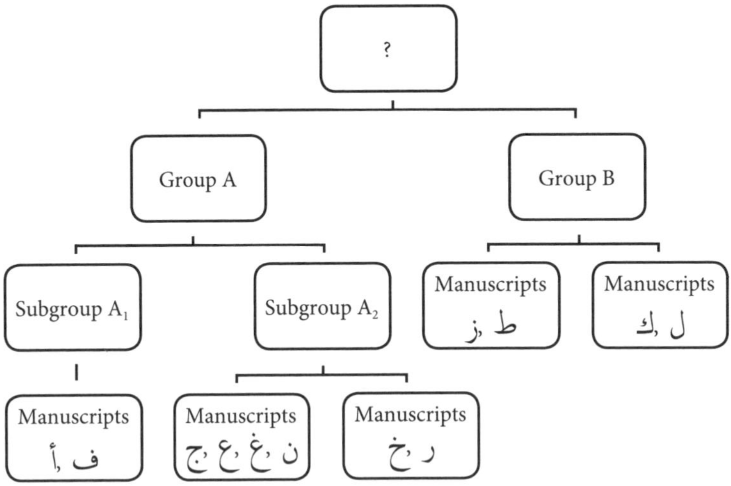

Figure 1. Textual Relationships

1.2. Edition

The present edition uses the text preserved in Esad Efendi, MS 3638 [أ] as a base text. This manuscript has been collated with the other eleven copies, and the omissions, additions, and most important textual discrepancies have been noted in the apparatus. This edition does not seek to reconstruct a hypothetical *Urtext* (if it ever existed), but rather to make available to the reader a reliable version of the epistle that takes into consideration the peculiarities of its transmission. The general tendency has been to preserve the version in manuscript أ, although occasional editorial interventions have opted for alternative readings, usually shared by several manuscripts. The divergences between the manuscripts as regards the main text are mainly orthographic and stylistic and do not hinder the comprehension of the epistle. The few relevant discrepancies affect the numerical values expressing length, time, longitude, and latitude. Manuscript أ is usually the most reliable in this regard, but, for the sake of consistency, preference has also been given to this copy even when other values appeared to be more plausible, unless the variants were shared by a relevant number of manuscripts. In all cases these variants have been preserved in the apparatus.

1.2.1. Apparatus

1.2.1.1. Omissions and additions

All the manuscripts present incidental omissions and additions that do not alter the meaning of the text, such as the omission of units of measurement after the quantity (e.g., *mayl, farsakh, sāʿa, daqīqa*), prepositions (notably, the compound preposition *min naḥw*), copulative conjunctions (especially the conjunction *thumma*), and variations in the formulaic expressions after the name of God and in the vocative formulas addressing the reader. Variants of this kind are not noted in the apparatus. Relevant omissions and additions include the spaces intentionally left blank by the copyists, notably when a numerical value is lacking, and other lacunae, additions or mistakes, that alter the meaning of the text. In this case, the omission is noted in the apparatus using the following notation:

Additions: في زيادة

Omissions: من سقط

1.2.1.2. Divergences

Although the majority of the variants do not affect the meaning of the text, these twelve manuscripts contain copious divergences. The following, which are the most frequent, have not been noted in the apparatus:

Use of the conjunction ف (*fa-*) instead of و (*wa*).

Use of variants of demonstrative pronouns: ذاك (*dhāka*) instead of ذلك (*dhālika*), and ذا (*dhā*) instead of هذا (*hādhā*).

Use of different plural forms, such as علّات (*ʿillāt*) instead of علل (*ʿilal*), أبحر (*abḥur*) instead of بحار (*biḥār*), بلاد (*bilād*) instead of بلدان (*buldān*), أجبل (*ajbul*) instead of جبال (*jibāl*).

Use of different connectors: وذلك (*wa dhālika*) instead of و (*wa*), فأنّ (*fa-anna*) instead of ف (*fa-*), بأنّ (*bi-anna*) instead of أنّ (*anna*), alone or with the annexed pronouns.

Another discrepancy that should be mentioned concerns the alteration of the order of the elements in the copious lists of the epistle. These permutations are relevant as regards the geographical location of the cities and lands listed in the description of each clime, and they are always noted in the apparatus.

1.2.1.3. Numerals

The manuscripts all use three different ways to express numerical values:

a. Numerals expressed in written form are employed particularly when referring to the number of cities and geographical features in each clime, and to express time and degrees. The rendition of cardinal numerals is extremely varied; there are discrepancies in the orthography, gender, concordance, and case. In this edition, the written form of the numbers is always normalized, and the divergences are not noted in the apparatus.

b. Arabic numerals are preferred for expressing long distances, such as the diameter of the Earth and the lengths of the climes, although they can also be used to express time, longitudes, and latitudes. It is worth noting that the form of the Arabic numerals is occasionally inconsistent in the same manuscript; for instance, the number 'four' occurs as ٤ and ۴, and the number 'zero' can be written with a simple point ٠, and also with a rounded shape ٥ that can be confused with the number 5, when written as ۵ or ٥.[1] Arabic numerals are often distinguished by a different colour of ink or a line above them. In this edition, these numerals are rendered using the standard Arabic forms.

c. *Abjad* numerals are sometimes used to express co-ordinates in the main text, and are always used in the tables. The *ductus* of the *abjad* numerals is usually difficult to read, and the majority of these letters do not bear any diacritical points — a fact that renders their comprehension very difficult and sometimes entirely conjectural.

1 The confusion of the Arabic numbers 'zero' and 'five' is a frequent mistake in previous editions of this epistle, such as the editions of Bombay (1887–1889), Cairo (1928), and Beirut (1955). On these editions, see I. K. Poonawala, 'Why We Need an Arabic Critical Edition', in *The Ikhwān al-Ṣafā' and their* Rasā'il. *An Introduction*, ed. N. El-Bizri (Oxford: OUP–IIS, 2008), pp. 33–57.

1.2.1.4. Formulaic Expressions

The manuscripts show a considerable diversity of eulogistic expressions when addressing God and the readers of the epistle, which have not been noted in the apparatus. The vocative expressions addressing the reader alternate between the simple formulas: أيها الأخ and يا أخي; and the most extended:

أيّها الأخ البارّ الرحيم (ayyuhā al-akh al-bārr al-raḥīm)

أيّها الأخ (يا أخي) البارّ الرحيم أيّدك الله وإيّانا بروح منه (ayyuhā al-akh [yā akhī] al-bārr al-raḥīm ayyadak Allāh, wa iyyānā bi-rūḥ minhu)

In this, as well as in the formulas invoking God's name, this edition follows the text of the base manuscript Esad Efendi, MS 3638 [أ]. The common formulas following the name of God are:

جل ثناؤه – عز وجل – سبحانه – تعالى (jall thanā'ūhu – ʿazz wa jall – subḥānhu – taʿālā)

1.2.1.5. Figures and Tables

The description of the Earth and the climes is accompanied by a series of images and tables that varies from manuscript to manuscript. The most common presentation of this cartographical information consists of two images representing the spherical Earth: the first one with a depiction of the equatorial line and the poles and/or cardinal points; the second one with a depiction of the seven climes; and eight tables containing the co-ordinates of important cities — the first table usually contains a general list of large cities, and the rest of the tables list the relevant cities in each clime. As mentioned earlier in the description of the manuscripts, manuscript غ contains an additional table with the most relevant Iranian cities.

The divergences in the layout of the tables and the variants concerning their contents are far too many to be noted in the apparatus, rendering it impossible to elaborate a critical reconstruction. We have opted for presenting an edition of the Arabic text of all the tables individually as an appendix.

1.2.1.6. *Paratexts*

1.2.1.6.1. Section Divisions

Mediaeval Arabic manuscripts use different methods to indicate the divisions in the text. The most usual practices in the case of the copies of the epistle 'On Geography' are:

Use of a different colour ink to mark the beginning of a section (ر, ج، خ، غ، ك، ل).

Use of overlined words to mark the beginning of a section (ز، ن).

Use of symbols such as ∴ or • to mark the end of a section (أ، خ، غ).

Use of rubrics with the title of the section, the generic term *faṣl* (section), usually written in a different ink. All the copies of the epistle use both kinds of rubric.

1.2.1.6.2. Titles and Closing Formulae

The manuscripts exhibit great divergence with regard to the title of the epistle. The variants are mainly formal and stylistic, but the numbering of the epistle varies. These discrepancies have not been indicated in the apparatus, but all the titles and endings are given below.

Atif Efendi, MS 1681 (ʿayn, ع)

Title:

الرسالة الخامسة في جغرافيا من القسم الأوّل من الرياضيات من جملة إحدى وخمسين رسالة لإخوان الصفاء في تهذيب النفس وإصلاح الاخلاق من كلام الصوفيّة

al-risāla al-khāmisa fī jughrāfiyā min al-qism al-awwal min al-riyyāḍiyāt min jumlat iḥdā wa khamsīn risāla li-ikhwān al-ṣafāʾ fī tahdhīb al-nafs wa iṣlāḥ al-akhlāq min kalām al-ṣūfiyya

Ending:

تمّت الرسالة بقوّة اللّه وتأييده

tammat al-risāla bi-quwwat Allāh wa taʾyīdihi

Esad Efendi, MS 3638 (alif, أ)

Title:

رسالة جغرافيا يعني صورة الأرض والأقاليم وهي السادسة من القسم الأوّل من الرياضيّات

risālat jughrāfiyā ya'anī ṣūrat al-arḍ wa'l-aqālīm wa hiyya al-sādisa min al-qism al-awwal min al-riyyāḍiyyāt

Ending:

تمّت رسالة جغرافيا وهي الخامسة (*sic*) والحمد لله حق حمده والصلوة على نبيّنا محمّد وآله وصحبه

tammat risālat jughrāfiyā wa hiyya al-khāmisa (*sic*) wa'l-ḥamd li-Allāh ḥaqq ḥamduhu wa'l-ṣalāt 'alā nabiyyinā Muḥammad wa ālihi wa suḥbihi

Esad Efendi, MS 3637 (nūn, ن)

Title:

الرسالة الخامسة في جغرافيا يعني صورة الأرض والأقاليم والبيان وأنّها كريّة الشكل بجميع ما عليها من البحار والجبال والبراريّ والأنهار والمدن والقرى وكيفيّة خططها ومسالكها وممالكها والغرض منها هو التنبيه على علّة ورود النفس إلى هذا العالم والحثّ على التفكّر في الأفاق وفي أنفس الغافلين عنها حتّى يتبيّن لهم الحقّ فيستعدون للرحلة والتزوّد إلى دار الآخرة قبل الممات وفاء العمر وتقارب الاجل والقوت والندامة

al-risāla al-khāmisa (sic) fī jughrāfiyā ya'anī ṣūrat al-arḍ wa'l-aqālīm wa'l-bayyān wa annaha kuriyyat al-shakl bi-jamī' mā 'alayhā min al-biḥār wa'l-jibāl wa'l-barāri wa'l-anhār wa'l-mudun wa'l-qurā wa kayfiyyat khiṭaṭhā wa masālikhā wa mamālikhā wa'l-gharḍ minhā huwa al-tanbīh 'alā 'illat wurūd al-nafs ilā hādhā'l-ālam wa'l-ḥathth 'alā'l-tafakkur fī'l-āfāq wa fī anfus al-ghāfilīn 'anhā ḥattā yatabayyin lahum al-ḥaqq fa-yasta'idūn li'l-riḥla wa'l-tazawwud ilā dār al-ākhira qabl al-mamāt wafā' al-'umr wa taqārub al-ajl wa'l-qawt wa'l-nadāma

Ending:

تمّت

tammat

Köprülü, MS 870 (kāf, ك)

Title:

[بياض] في جغرافيا يعني صورة الأرض والأقاليم من رسائل إخوان الصفاء صان الله أقدارهم بسم الله الرحمن الرحيم

[bayāḍ] fī jughrāfiyā ya'nī ṣūrat al-arḍ wa'l-aqālīm min rasā'il ikhwān al-ṣafā' ṣān Allāh aqdārhum b'ism Allāh al-raḥmān al-raḥīm

Ending:

تمّت الرسالة الرابعة الجغرافيا في صورة الأرض وهئتها

tammat al-risāla al-rābi'a al-jughrāfiyā fī ṣūrat al-arḍ wa hay'atihā

Köprülü, MS 871 (lām, ل)

Title:

الرسالة الرابعة في جغرافيا يعني صورة الأرض والأقاليم والبيان فإنّها كريّة الشكل [في الهامش: من رسائل إخوان الصفاء صان الله أقدارهم]

al-risāla al-rābi'a fī jughrāfiyā ya'nī ṣūrat al-arḍ wa'l-aqālīm wa'l-bayyān fa-innahā kuriyyat al-shakl [fī al-hāmish: min rasā'il ikhwān al-ṣafā' ṣān Allāh aqdārhum]

Ending:

تمّت الرسالة والحمد لله كما هو أهله ومحقّه وصلى الله على سيّدنا محمّد النبيّ وآله الطيّبين الطاهرين

tammat al-risāla wa'l-ḥamd li-Allāh kamā huwa ahlhu wa muḥiqqhu wa ṣalā Allāh 'alā sayyidnā Muḥammad al-Nabī wa ālihi al-ṭayyibīn al-ṭahirīn

Feyzullah, MS 2130 (fā', ف)

Title:

الرسالة السادسة في جغرافيا وهي صورة الأرض والأقاليم وأنّها كريّة الشكل بجميع ما عليها من الجبال والبحار والأنهار والبراريّ والمدن والقرى وكيفيّة خططها ومسالكها وممالكها والغرض منها هو التنبيه عن ورود النفس إلى هذا العالم لتستعد للرحلة والتزوّد إلى دار الآخرة قبل الممات وفاء العمر وانفضاء الاجل ووجدان الجسدة والندامة

al-risāla al-sādisa fī jughrāfiyā wa hiya ṣūrat al-arḍ wa'l-aqālīm wa annahu kuriyyat al-shakl bi-jamī' ma 'alayhā min al-jibāl wa'l-biḥār wa'l-anhār wa'l-barārī wa'l-mudun wa'l-qurā wa kayfiyyat khiṭaṭhā wa masālikhā wa mamālikhā wa'l-gharḍ minhā huwa al-tanbīh 'an wurūd al-nafs ilā hādhā al-'alam li-tasta'idd li'l-riḥla wa'l-tazawwud ilā dār al-ākhira qabl al-mamāt wafā' al-'umr wa infiḍā' al-ajl wa wijdān al-jisda wa'l-nadāma

Ending:

تمّت الرسالة السادسة الموسومة بالجغرافيا أي صورة الأرض من القسم الأوّل

tammat al-risāla al-sādisa al-mawsūma bi'l-jughrāfiyā ayy ṣūrat al-arḍ min al-qism al-awwal

Tehran, MS 7437 (ṭāʿ, ط)

Title:

الرسالة الرابعة في جغرافيا يعني صورة الأرض من رسائل إخوان الپعاء (*sic*) صان اللّه أقدارهم

al-risāla al-rābiʿa fī jughrāfiyā yaʿnī ṣūrat al-arḍ min rasāʾil ikhwān al-paʿāʾ (sic) ṣān Allāh aqdārhum

Ending:

تمّت الرسالة الرابعة في جغرافيا من رسائل إخوان الپعاء (*sic*)

tammat al-risāla al-rābiʿa fī jughrāfiyā min rasāʾil ikhwān al-paʿāʾ (sic)

Bibliothèque nationale, MS 2303 (rāʿ, ر)

Title:

هذا رسالة في جغرافيا وهي الرسالة الخامسة من القسم الأوّل الرياضيّ من كتاب رسائل إخوان الصفاء أيّدهم اللّه تعالى بروح منه أمين

hādhā risāla fī jughrāfiyā wa hiya al-risāla al-khāmisa min al-qism al-awwal al-riyāḍī min kitāb rasāʾil ikhwān al-ṣafāʾ ayyadhum Allāh taʿālā bi-rūḥ minhu amīn

Ending:

تمّت رسالة جغرافيا والحمد للّه ربّ العالمين

tammat risālat jughrāfiyā wa'l-ḥamd li-Allāh rabb al-ʿālamīn

Bibliothèque nationale, MS 2304 (zayʿ, ز)

Title:

الرسالة الرابعة في جغرافيا يعني صورة الأرض والأقاليم من رسائل إخوان الصفاء

al-risāla al-rābiʿa fī jughrāfiyā yaʿnī ṣūrat al-arḍ wa'l-aqālīm min rasāʾil ikhwān al-ṣafāʾ

Ending:

تمّت الرسالة الرابعة الجغرافيا في صورة الأرض وهيئتها

tammat al-risāla al-rābiʿa al-jughrāfiyā fī ṣūrat al-arḍ wa hayʾathā

Bodleian Library, Oxford, MS Hunt 296 (jīm, ج)

Title:

الرسالة الجغرافيا في صورة الأرض والأقاليم وهي الخامسة من القسم الأوّل من الرياضيّات العلميّة

al-risāla al-jughrāfiyā fī ṣūrat al-arḍ wa'l-aqālīm wa hiya al-khāmisa min al-qism al-awwal min al-riyāḍiyyāt al-ʿilmiyya

Ending:

تمّت رسالة جغرافيا وهي الخامسة من رسائل إخوان الصفاء في تهذيب النفس والحمد للّه وحده

tammat risālat jughrāfiyā wa hiya al-khāmisa min rasāʾil ikhwān al-ṣafāʾ fī tahdhīb al-nafs wa'l-ḥamd li-Allāh waḥdihi

Bodleian Library, Oxford, MS Laud Or. 260 (ḥāʾ, خ)

Title:

رسالة جغرافيا يعني صورة الأرض والأقاليم وهي الخامسة من الرياضيّات من القسم الأوّل

risālat jughrāfiyā yaʿnī ṣūrat al-arḍ wa'l-aqālīm wa hiya al-khāmisa min al-riyāḍiyyāt min al-qism al-awwal

Ending:

تمّت رسالة جغرافيا وهي الخامسة من الرياضيّات من القسم الأوّل بحمد للّه ومنّته وصلواته على سيّدنا محمّد المصطفى وآله وصحبه أجمعين الطاهرين

tammat risālat jughrāfiyā wa hiya al-khāmisa min al-riyāḍiyyāt min al-qism al-awwal bi-ḥamd Allāh wa minnatihi wa ṣalawātihi ʿalā sayyidnā Muḥammad al-muṣṭafā wa ālihi wa ṣuḥbihi ajmaʿīn al-ṭāhirīn

Bodleian Library, Oxford, MS Marsh 189 (ghayn, غ)

Title:

كتاب جغرافيا وهو الخامس في صورة الأرض والأقاليم

kitāb jughrāfiyā wa huwa al-khāmis fī ṣūrat al-arḍ wa'l-aqālīm

(No indication of ending)

1.2.2. Editorial Interference

1.2.2.1. Normalization of Orthography and Syntax

Orthography and syntax exhibit a great degree of diversity and have been normalized in the edition. The principal editorial intervention affects the following aspects:

Reconstruction of the *scriptura plena* in those cases in which the scribes have always copied the *ductus* without diacritical points; this affects especially the toponyms and the *abjad* numerals.

Occasional addition of vowels to clarify the meaning.

Normalization of the *alif*, for instance: هكذا for هكذى

Normalization of the medial and final *hamza*, for instance: حكماء for حكما ;أجزاؤها for أجزاها ; or ماية for مائة

Normalization of participles of defective verbs, for instance: ناتٍ for ناتي

Gender and numerical concordance between subjects and verbs.

Normalization of gender concordance of relative pronouns.

Orthography and concordance of numerals.

1.3. Translation

The epistle 'On Geography' of the Ikhwān al-Ṣafā' has been translated several times. Its first translation was into Latin, with the title *Epistola Fratrum Sincerorum in Cosmographia*. This epistle, as well as cognate works such as those of al-Farghānī or al-Battānī, represents an oddity in the corpus of Latin translations of Arabic works, where geography was clearly under-represented. The translation, which was probably made in Italy in the thirteenth or fourteenth century, is very close to the original Arabic text, although the Qur'anic references have been expunged and the section containing the allegory of the wise king is missing. The figures given in this translation for the lengths of the circumference of the Earth and each one of the seven climes do not correspond to those in any of the manuscripts consulted for our edition.

There are also two modern translations of the epistle 'On Geography'. The oldest one is a partial German translation made by Friedrich Dieterici. This translation also includes the tables, which are presented as an appendix, but several passages that the translator has judged either not relevant or not proper to the discipline of geography, such as the proemium, the allegory of the wise king, and the final discussion concerning the cycles of time, have been paraphrased or simply ignored; the Qur'anic quotations have also been omitted.

Alessandro Bausani published a partial Italian translation of the entire collection of Epistles in 1978. However, rather than being a partial translation like the German, this version is a summarized paraphrase, namely a selection of the passages that the translator has considered as being properly related to the subject of the treatise, i.e., geography. The admonitory sections addressed to the reader, the proemium, the allegory of the wise king, and the concluding considerations discussing the cycles of time were ignored, as were the Qur'anic quotations.

The need for a new translation of the epistle 'On Geography' is the obvious consequence of the new critical edition, as none of the previous translators had the opportunity of using a text based on a critical comparison of twelve manuscripts. But it is also an opportunity to present the treatise of the Ikhwān in a different light. The omission in previous translations of critical parts of the text and of the Qur'anic quotations substantially alters the meaning of the epistle. By eliminating these non-technical sections, not only did the translators cover the treatise with a patina of anachronistic scientism, but they also deprived the epistle of its intellectual and spiritual context.

The principal aim of this new translation has been to provide a readable version close to the Arabic text, accompanied by an explanatory apparatus to facilitate the reader's comprehension. This translation presents the entire text of the epistle in English for the first time, and it is the first annotated version. The text is repetitive, both in its technical and hortatory sections; it contains extended enumerations, and, like so many Arabic treatises from this epoch, abounds in long and convoluted sentences. This translation has tried to preserve the original syntax as much as possible, and also to follow the original text closely as far as the repetitions are concerned since the syntax is highly idiosyncratic and is more representative of the style of the anonymous authors than any other feature of the epistle.

Typographical decisions have also been made on the basis of keeping the translated text as close as possible to the original. In this regard, and although the choice could be regarded as counter-intuitive, all the enumerations have been presented in running text. Likewise, the rendition of the numerical data concerning length, time, and latitude and longitude, has followed the original text, and the figures may occur as numbers or written numerals accordingly; in the case of the degrees, however, the figures have been always normalized and added within the running text between brackets. A further licence concerns the division of the text into paragraphs, which do not always correspond to the sections of the Arabic.

The extreme inconsistency of the accompanying tables, which list the most important cities and their co-ordinates, renders it impossible to produce a critical edition in the true sense. They have been edited separately and presented as an appendix as they appear in the manuscripts. Both the diagrams and the tables used for the translation correspond to manuscript أ. Albeit incomplete, this manuscript contains one of the most consistent versions of the tables, and it has been considered preferable to translate them, rather than to present an artificial reconstruction based on all the copies. A comparative list of all the cities mentioned in the tables of the manuscripts has been added as an appendix, as well as a comparison of the — often miscopied — co-ordinates, which have been collated, whenever possible, with the co-ordinates provided by al-Khwārizmī in his *Kitāb Ṣūrat al-arḍ*.

The text contains numerous toponyms and ethnonyms. The preferred option has been always to use the normalized version of the name whenever possible; the names that do not have a standard version in English have been transliterated. Although the notes occasionally may convey short explanatory texts, they have been mainly used to provide information about the locations mentioned in the text and to note the variants concerning the length and position of the climes. The English translation of the Qur'anic quotations has been taken from Arberry's *The Koran Interpreted*.[1]

1 A. J. Arberry, *The Koran Interpreted* (London: Allen & Unwin, 1955).

1.4. Geography

1.4.1 Arabic Geography in Mediaeval Islam

Geographical and cartographical knowledge occupies a central place in classical Islamic civilization. The Arabic term for geography, *jughrāfiyā*, was directly borrowed from Greek, and was often translated into Arabic as *ṣūrat al-arḍ* and used by early Muslim geographers as the title of several treatises. The science of geography in Islam, however, did not develop solely on Hellenistic scientific foundations, but also on those laid by other cultures of the Antique world, notably the Iranian and the Indian traditions.

From the body of geographical literature produced by Muslim authors in the first centuries of Islam, approximately two dozen works have survived.[1] The characterization of these treatises as 'geography' is by no means unproblematic. The scope of these works usually goes far beyond the limits of the Greek geographical tradition *stricto sensu*, as they were often envisioned as an all-encompassing — and sometimes disorganized — description of the world that included, on the one hand, information about itineraries and tax duties, geographical features, location co-ordinates, and toponymical etymology; and, on the other hand, historical and anthropological information, poetry and anecdotes of ethical and literary value, and descriptions of various wonders. Despite this conventional classification of the contents into technical and non-technical, which has been widely used in classical studies on this subject to demarcate the boundaries of a proper geographical discipline,[2] it is not possible to set a generic demarcation between physical and human geography in early Muslim geographical writings. So, too, it is not possible to envisage a purely scientific approach in an intellectual environment where science was regarded as a way of comprehending God's designs.

---

1 For a succinct survey of these early sources, see A. Silverstein, 'The Medieval Islamic Worldview: Arabic Geography in its Historical Context', in *Geography and Ethnography: Perceptions of the World in Pre-Modern Societies*, ed. K. A. Raaflaub and R. J. A. Talbert (Chichester: Wiley–Blackwell, 2010), pp. 273–90.

2 Cf. for instance, A. Miquel, *La géographie humaine du monde musulman jusqu'au milieu du 11e siècle*, 4 vols. (Paris: Éditions de l'EHESS, 1973–1988).

The use of this and similar dichotomies is often an obstacle to the proper understanding of Islamic geographical works, which in the first centuries of Islam were more than an endeavour to describe the visible world using scientific methods.[1] Regardless of the scope and the goals of each particular work, the writings of the Muslim geographers were deeply imbued with religious meaning, and their treatises should be understood as a comprehensive attempt to describe Creation in its entirety and, consequently, as means of praising its Creator. We should also take into account the point that the faith these works aimed to serve was neither a hypostatized entity, nor the abstract and specialized religion of the scholars, but the Islam of the empire. The discipline of geography was fostered predominantly according to the needs of the centralized ʿAbbāsid state and it therefore received direct caliphal support, as did other disciplines that helped to shape ideology. As Paul Heck has shown, discourse on Islam as a state and discourse on Islam as a religion occupied the same epistemological space articulated in a single language.[2]

In terms of their cosmological framework and multi-disciplinary nature, and despite the abstraction of the most technical works, such as those on cartography, early Islamic geographical writings were a confirmation of the pre-eminence of Islam. The work of the Muslim cosmographer was an act of worship: it entailed the realization of the divine order of the physical world and also of the infinite variety of God's Creation. It was, at the same time, an act of political significance, since the celebration of God's Creation was consubstantial to the celebration of the divine mission of the Caliph, God's vicegerent on Earth and, very often, the sponsor of these works.[3] Lastly, and due to the scope and breadth of the endeavour and its scientific achievements, it was furthermore a vindication of the pre-eminence of Islam over

1 See the discussion of the technical/non-technical dichotomy and the study of geography vis-à-vis *adab* in J. Montgomery, 'Serendipity, Resistance, and Multivalency', in *On Fiction and* Adab *in Medieval Arabic Literature*, ed. P. F. Kennedy (Wiesbaden: Harrassowitz, 2005), pp. 177–230.

2 Paul L. Heck, *The Construction of Knowledge in Islamic Civilization* (Leiden: Brill, 2002), p. 3.

3 Montgomery, 'Serendipity, Resistance, and Multivalency', p. 214.

superseded forms of religion and philosophy, especially the Greek tradition.[1]

The multivalency of these treatises and the extraordinary flowering of this genre in the early ʿAbbāsid period cannot be understood, therefore, without taking into consideration the relationship between religion and pre-modern science, the wider cultural context of the Late Antique world to which the Middle East belonged, and the ideological motives behind the cultural policies of the ʿAbbāsid caliphs.

This is the background against which the Ikhwān's epistle 'On Geography' should be read. Despite all the ways in which it differs from many of the previous examples of geographical literature, it shares their polysemous nature, being both a product of the Hellenistic scientific tradition and a work profoundly rooted in Islamic tenets; and it is also a vindication of a particular conception of Islam, whose characteristics should be looked for in the entire collection of epistles. In this regard, and unlike earlier ʿAbbāsid treatises, the anonymity of the project and the ambiguous relationship of its authors with Ismailism compels the reader to consider the possible political agenda that is likely to have inspired the composition of these epistles, and to examine the relationship between ʿAbbāsid authors and authorities in a different light.

1.4.2. The Ptolemaic Tradition of Mathematical Geography

The history of the first centuries of Islam, including historical continuity in the intellectual and scientific fields, is intimately connected with the Late Antique world, to the extent that scholars have started to use the term 'Islamic Late Antiquity' to describe the period that lasted from the Muslim conquests of the Middle East until the fourth/tenth century.[2] The Hellenistic influence was predominant in all fields of science, with the Greek geographical tradition playing a central role in the development of Islamic geography. However, it was by no means the only source available to the Muslim geographers; numerous important

1 Ibid., p. 181.
2 See, for instance, T. Sizgorich, 'Narrative and Community in Islamic Late Antiquity', *Past & Present*, vol. 185 (2004), pp. 9–42.

geographical concepts were borrowed from the Indian and Iranian traditions. The idea of the cupola of the Earth, which was situated in Sri Lanka (*Sirindīb*) and from where longitudes were calculated, and the belief that the inhabited regions of the world were only in the Northern Hemisphere, for instance, were concepts that some Muslim geographers took from Indian geography. Among the Iranian influences that permeated Muslim geographical works, the most important is the adoption by many geographers of the sevenfold organization of the world into *kishwār*s (lands, regions), and the Iranian location of the *omphalos mundi* in the Sawād (the fertile lands of southern Iraq), conceived of as the centre of the Iranian empire or 'Greater Iran' (*dil-i īrān-shahr*) during the Sassanid era, a conception which has clear ideological overtones in the Islamic period.[1] For all the importance of these contributions, the major influence in the development of Islamic geography is indubitably Hellenistic, particularly the works of Ptolemy which were translated into Arabic at an early date.

The translation movement initiated under the aegis of the first ʿAbbāsid caliphs, and especially the caliph al-Maʾmūn (198–218/813–833), is one of the greatest intellectual achievements in the history of Islam. By the end of the fourth/tenth century, almost all the Greek scientific and philosophical works available throughout the Byzantine Empire and the Near East had been translated into Arabic.[2] It seems that the geographical works of Ptolemy, or at least parts of them, were circulated in Arabic translation as early as the second half of the second/eighth century;[3] but the real origin of Islamic mathematical geography should be sought in the House of Wisdom (*bayt al-ḥikma*), the caliphal library sponsored by al-Maʾmūn. By the beginning of the third/ninth century, al-Maʾmūn had commissioned a group of geographers and

1 On these influences, cf. J. H. Kramers, 'L'influence de la tradition iranienne dans la géographie arabe', in his *Analecta Orientalia* (Leiden: Brill, 1954), vol. 1, pp. 147–56; and S. Maqbul Ahmad, 'Djughrāfiyā', in *EI2*, vol. 2, p. 576, where he discusses the existence of two schools of descriptive geography: Iranian and Balkhī.

2 On this process, see D. Gutas, *Greek Thought, Arabic Culture* (London: Routledge, 1998).

3 Cf. F. Sezgin, *Geschichte des arabischen Schrifttums. Mathematische Geographie und Kartographie im Islam*, vol. 10 (Frankfurt: Institut für Geschichte der Arabisch-Islamischen Wissenschaften, 2000), p. 79.

astronomers to create a world map. According to al-Masʿūdī, the Maʾmūnid map (*al-ṣūra al-maʾmūniyya*) depicted 'the universe with its spheres, stars, lands and seas; its uninhabited and inhabited parts, settlements of peoples, cities, and similar things', and was better than previous maps, such as those of Marinus of Tyre and Ptolemy.[1]

This map was the culmination of an ambitious scientific endeavour that brought together some of the most prominent Muslim scholars. The project entailed the translation of the classical geographical and astronomical sources, but also their emendation with new calculations and data collected through direct observation. Ptolemy was certainly the most important source. His *Almagest* (*al-Majisṭī*) was rendered into Arabic by Yaḥyā ibn Khālid ibn Barmak (d. 191/806–807) and was the subject of numerous commentaries.[2] Ptolemy's cartography was also translated in the late second/eighth or early third/ninth century, although the exact date is uncertain; we know that al-Kindī (d. 260/874) made a translation of this work, but there were already previous versions, as is shown in the works of two of the scholars who took part in al-Maʾmūn's project and who, as we shall discuss, are the two main sources for the epistle 'On Geography' of the Ikhwān al-Ṣafāʾ: al-Khwārizmī (d. 235/850) and al-Farghānī (third/ninth century).

These authors adopted and systematized the main scientific concepts of the Hellenistic tradition, establishing the Ptolemaic framework that would be used in later geographical and cartographical Arabic writings. From Ptolemy and other Greek authors were borrowed the main concepts that inform the thought of the early ʿAbbāsid geographers and, to a great extent, of the Ikhwān al-Ṣafāʾ as well: the *oekumene*, the inhabited land which corresponds to a quarter of the Earth (*al-rubʿ al-maskūn min al-arḍ*); the division of the Northern Hemisphere

1 ʿAlī ibn al-Ḥusayn al-Masʿūdī, *Kitāb al-Tanbīh waʾl-ishrāf* (Leiden: Brill, 1894, rpr., 1967), p. 33; Sezgin, *Geschichte des arabischen Schrifttums*, vol. 10, p. 81. The Map of al-Maʾmūn has not come down to us, but al-Zuhrī (fl. 532/1137) based his *Kitāb al-Juʿrāfiyā* on a copy of this map made by al-Fazārī. Since al-Zuhrī's arrangement is based on the Iranian system of *kishwar*s, it is possible that al-Maʾmūn's map represented a synthesis of Iranian and Ptolemaic traditions; cf. S. Maqbul Ahmad, 'K͟harīṭa', in *EI2*, vol. 4, p. 1078 f.

2 See C. A. Nallino, 'Al-Ḫuwārizmī e il suo rifacimento della Geografia di Tolomeo', in *Atti della Real Accademia dei Lincei*, anno 291, serie V, *Memorie delle Classe di Scienze Morali* (1894), vol. 2, pp. 4–53, on p. 6.

into seven latitudinal climes (*aqālīm*); the expression of geographical location in terms of longitude (*ṭūl*) and latitude (*ʿarḍ*); the idea of the four continents (Europe, Libya, Ethiopia, and Scythia); the belief that the Indian Ocean was a landlocked sea; and, of course, the name of the discipline, *jughrāfiyā*.[1]

1.4.3. Contents and sources of the epistle 'On Geography' of the Ikhwān al-Ṣafāʾ

The contents of this epistle can be summarized in the following points:

a) Proemium: geography as one of the noble sciences (*al-ʿulūm al-sharīfa*) [pp. 49–50]
b) Description of the Earth, the equator and the cardinal points (with diagrams of the Earth) [pp. 50–57, 61–62]
c) Discussion of common beliefs concerning the highest and lowest points [pp. 51–52]
d) Discussion of the sphericity of the Earth [pp. 50–52]
e) Discussion of the four theories explaining the immobility of the Earth [pp. 52–53]
f) Description of the shape of the Earth [pp. 50–53]
g) Description of the inhabited quarter of the Earth [pp. 55–56]
h) Boundaries of the seven climes and the origin of these divisions [pp. 57–59, 61]
i) Allegory of the city of the wise king and the value of scientific knowledge [pp. 59–60]
j) Geographical features, lands, and cities of the seven climes (with cartographical tables) [pp. 57–75]
k) Section on the influence of the stars, cosmic dualism, and the opposition between the virtuous people [*ahl al-khayr*] and the evil-doers [*ahl al-sharr*] [pp. 75–77]

It has been suggested that the epistle 'On Geography' of the Ikhwān al-Ṣafāʾ was probably the first tract that used the Greek loan word *jughrāfiyā* to refer to 'a map of the world and the climes'.[2] The only

1 On the Greek influence, see Ahmad, 'D̲j̲ug̲h̲rāfiyā', in *EI2*, vol. 2, p. 577.
2 Ibid., p. 575.

direct reference that we find in the text, which occurs in the context of the discussion of variations in the length of the day in the different climes, is to Ptolemy's *Almagest*,[1] and the Ptolemaic influence is evident in the use of both astronomical and cartographical concepts such as the practice of calculating the longitudes from a prime meridian located in the extreme West, and the methodology employed to locate the centre and the latitudinal boundaries of the climes. Also Hellenistic, though not exclusively Ptolemaic, is the idea of the *oekumene* and its division into seven climes, which provides the basic structure for the epistle. In contrast, the Greek division of the inhabited world into four provinces (Europe, Libya, Ethiopia, and Scythia), which was used by other Muslim authors, is not taken into consideration in the Ikhwān's description of the world.

However, this is not a work directly inspired by Ptolemy. Its authors drew upon previous efforts to recover and improve the studies on the Greek author, and it is possible to identify the main sources used by the Ikhwān: the particular translation of Ptolemy's *Almagest* on which they relied is unknown, but the information deriving from Ptolemy's *Geography* was collected from the treatises of al-Khwārizmī and al-Farghānī. The epistle 'On Geography' is not a mere recension of Ptolemaic translations or adaptations either; the imprint of Iranian geography is also of great significance and has clear ideological implications, and the plan of the treatise is determined by the Muslim Neoplatonism that inspires the epistles in their entirety.

The description of the Earth begins with a discussion of its position in the centre of the celestial sphere and its immobility. Although the Ikhwān do not discuss the movement of the stars in this epistle, they enumerate the four theories that were proposed to explain the immobile, central position of the Earth. The first two theories are clearly based on Ptolemaic celestial mechanics: (1) the Earth stays immobile in the centre of space due to the equal attractive force it experiences from the celestial sphere on all its points; or, conversely, (2) the Earth stays immobile due to the equal repulsive force from the planets on all its points. The third and fourth theories, in contrast, are based on the Aristotelian idea of 'natural place': (3) the centre of the Earth acts as a

1 Cf. below, p. 61.

magnet for the heavy bodies, which occupy a space around it according to the natural characteristics of the four universal components of the sub-lunar sphere which they possess, i.e., earth, water, air, and fire; the same principle determines the place of the planets around the Earth in the celestial sphere. This idea is dealt with by Ptolemy in the *Almagest*[1] and also used by the Ikhwān to explain the sphericity of the Earth, which is not discussed in this epistle in terms of Euclidean geometry. The last theory, which is the most plausible according to the authors, involves a religious interpretation of this Aristotelian idea: (4) the Earth stays immobile at the centre of the universe because this tendency of the elements towards their natural place accords with God's plan. He has determined the most appropriate place for each element in the Earth, as He has also done for the position of the planets in the celestial sphere.

The chorographical methodology used in this epistle to determine the limits of the seven climes in the inhabited quarter of the world is also inspired by Ptolemy's *Geography*. Each one of these seven climes has a central point where its longest day, on the summer solstice, lasts an exact number of half-hours in length; the length of the longest day in each clime increases by half an hour from clime to clime, thus ranging from the 13 hours of the first clime to the 16 hours of the seventh clime. The boundaries between climes are on the quarters, from the first boundary of the first clime, where Midsummer's Day has 12h 15', to the 16h 15' of the second boundary of the seventh clime.[2] The translation of these measures into degrees and units of length differs from the Greek tradition, but they correspond with the measurements provided by the intermediate source, al-Farghānī. Likewise, the tables with cities and co-ordinates that accompany the description of the climes are Ptolemaic in conception, as is to a great extent the nomenclature of the toponyms, but the direct source from which the Ikhwān collected their cartographical information was al-Khwārizmī.

1 Cf. O. Pedersen, *A Survey of the* Almagest (Odense: Odense University Press, 1974, rpr., New York: Springer Science, 2011), p. 43.

2 Cf. G. R. Tibbetts, 'The Beginnings of a Cartographical Tradition', in *History of Cartography*, ed. J. B. Harley and D. Woodward (Chicago: Chicago University Press, 1992), vol. 2, p. 102.

Al-Farghānī was one of the scholars involved in al-Maʾmūn's project to collect astronomical and geographical information and was one of the Muslim commentators on Ptolemy's *Almagest*. His most famous and important work is an astronomical treatise entitled *Jawāmiʿ ʿilm al-nujūm wa uṣūl al-ḥarakāt al-samāwiyya* (*Compendium of Astronomy and the Principles of Celestial Motions*). This compendium is an introduction to the principles of Ptolemaic astronomy in thirty chapters of which chapters six to nine, devoted to the description of the Earth, plausibly constitute the primary source used by the anonymous Ikhwān in their epistle 'On Geography'. These chapters of the *Jawāmiʿ* address the following topics:[1]

a) Chapter 6: on the location of the inhabited quarter of the Earth, the equatorial line, the poles and the cardinal points, and the differences between seasons.
b) Chapter 7: on the characteristics of the inhabited quarter and the places where the Sun rises and sets for several months.
c) Chapter 8: on the division of the inhabited quarter into seven climes, their boundaries, and the length of the circumference and diameter of the Earth.
d) Chapter 9: on the regions that each clime traverses and the most important cities in each.

It seems evident that these chapters inspired the epistle 'On Geography'. The topics treated are the same, and the epistle follows, for the most part, the order in which they are discussed by al-Farghānī. His influence is also evident in the factual information contained in this epistle. The lengths of the circumference of the Earth given in both treatises are the same: al-Farghānī calculates that the value of a degree on the equator corresponds to 56 ⅔ miles[2] and that the circumference of the Earth is

1 For a detailed description of all the chapters of this treatise, see B. Abdukhalimov, 'Aḥmad al-Farghānī and his *Compendium of Astronomy*', *Journal of Islamic Studies*, vol. 12, 2 (1999), pp. 142–58 and pp. 149–54.

2 Abu'l-ʿAbbās Aḥmad ibn Muḥammad al-Farghānī, *Kitāb Jawāmiʿ ʿilm al-nujūm*, edited and translated by J. Golius as *Elementa Astronomica* (Amsterdam, Jasonius à Waasberge: 1669), p. 30.

20,400 miles long,[1] which are the same figures given in the epistle.[2] A further corroboration can be found in the comparison of the sizes of the planes, as both treatises argue that the smallest star is eighteen times the size of the Earth.[3] As Gauthier Dalché has noted, when defining the concept of longitude, both works seem to place the prime meridian in the East (*min al-mashriq ilā al-maghrib*), as in the Indian tradition, although in the rest of these works the Greek practice of situating the prime meridian in the West is always followed.[4]

The latitudinal boundaries of the climes also correspond to the measurements of the *Jawāmiʿ* as can be seen in the following table:[5]

	Ikhwān	al-Farghānī	al-Battānī	al-Bīrūnī
	50° 30'	50° 30'	missing	50° 25'
Seventh Clime	48° 40' (or 48° 30')	48° 55'	48° 53'	48° 52'
	47°15'	47° 15'	47° 12'	47° 11'
Sixth Clime	45° 50'	45° 24'	45° 22'	45° 22'
	43° 30'	43° 30'	43° 35'	43° 23'
Fifth Clime	41° 20'	41° 20'	41° 15'	41° 14'
	39°	39° 00'	38° 54'	39° 00'
Fourth Clime	36° 50'	36° 24'	36° 22'	36° 21'
	33° 30'	33° 40'	33° 37'	33° 37'
Third Clime	30° 35'	30° 42'	30° 40'	30° 39'
	27° 30'	27° 30'	27° 28'	27° 28'
Second Clime	24° 06'	24° 06'	24° 05'	24° 04'
	20° 30'	20° 30'	20° 28'	20° 27'

1 Ibid., p. 31.

2 Manuscripts ج and خ give 30,400 miles, ع 24,000 miles, and ن 20,431 miles, but in all the manuscripts consulted for this edition the length of a degree on the equator is 19 *farsakh*s, and the diameter of the Earth is approximately 6,500 miles; these values correspond to a circumference of 20,400 miles.

3 Cf. Below, pp. 58; and al-Farghānī, *Elementa Astronomica*, (Arabic text) p. 85.

4 P. Gautier Dalché, '*Epistola Fratrum Sincerorum in Cosmographia*: Une traduction latine inédite de la quatrième *Risāla* des Ihwān al-Ṣafā'', *Revue d'histoire des texts*, vol. 18 (1988), p. 149.

5 I take the values of al-Farghānī, al-Battānī, and al-Bīrūnī from E. Honigmann, *Die sieben Klimata* (Heidelberg: Carl Winter, 1929), p. 163; see also Tibbetts, 'The Beginnings of a Cartographical Tradition', in *History of Cartography*, vol. 2, Fig. 4.6.

	Ikhwān	al-Farghānī	al-Battānī	al-Bīrūnī
First Clime	16° 40'	16° 40'	16° 39'	16° 39'
	12° 45'	missing	missing	12° 30'

Table 1 - Boundaries of the Climes Compared

The description of the climes given in the epistle 'On Geography' also bears a great resemblance to al-Farghānī's *Jawāmiʿ* in several aspects: the phraseology and the vocabulary, consisting of the repeated formula 'passes through' (*yamurr ʿalā*), is repeatedly used in both works; each clime is related to a planet at the beginning of the description; the enumeration of the regions which each clime traverses is very similar in both treatises; and many of the cities listed by al-Farghānī can be found, following the same order in the enumeration, in the tables of the epistle.[1]

Although al-Farghānī is the principal source for the main text, the composition of the cartographical tables drew on a different source. The tables of the cities with their co-ordinates that accompany the description of the seven climes are based on al-Khwārizmī's reformulation of Ptolemy's *Geography*. The extant manuscripts of the epistle 'On Geography' are extraordinarily dissimilar as regards the rendition of the names of the cities and their co-ordinates. These variants we may assume are most likely the result of scribal errors in the transmission of the foreign toponyms and ethnonyms, and in the rendering of the longitudes and latitudes, which are expressed in *abjad* numerals and therefore all too susceptible to error. Nevertheless, it is possible to discern the Khwārizmian origin of the chorographical information in the co-ordinates of the most accurate manuscripts — especially manuscript أ — and also in the occurrence of Ptolemaic toponyms whose Arabic transcription corresponds to the one used by al-Khwārizmī.[2]

There is, however, a striking difference that separates this epistle by the Ikhwān al-Ṣafāʾ from al-Farghānī's and al-Khwārizmī's works and

1 See Appendix 1.
2 See Appendix 2.

also from the majority of Islamic geographical writings: the treatment of the fourth clime.

1.4.4. The Fourth Clime: Geography and Prophetism

The standard description of the climes into which the inhabited world is divided consists of the determination of their latitudinal boundaries; the mention of the number of mountains, rivers and cities therein; the enumeration of the lands that they traverse; and the colour of their peoples. The fourth clime, however, includes an excursus explaining why it is the most excellent of all, and extolling the superior qualities of its inhabitants. This description seems to be inspired by the Iranian geographical tradition. Like authors of the Iraqi school such as al-Masʿūdī, al-Yaʿqūbī, and Ibn Khurradādhbih,[1] the Ikhwān al-Ṣafāʾ emphasize the excellence of the fourth clime due to its centrality in the septipartite clime system and the ascendancy that the Sun, the most important star, exerts over it.[2] Giving centrality to the Sawād or, specifically, to Baghdad, is a common trait in the geographical writings of the authors of this school. But this does not affect the narrative structure of the epistle 'On Geography', since the Ikhwān follow a strict geographical order, starting from the equatorial line. There is no explicit mention of Bābil or Baghdad either, as there is in al-Masʿūdī's *Kitāb al-Tanbīh*, or of the term 'heart of Iran' used by the Persian kings to refer to this region, as in Ibn Khurradādhbih's *Kitāb al-Masālik wa'l-mamālik*.

In this scheme, Mecca, the birthplace of the Prophet Muhammad, belongs to the second clime. Therefore to consider the fourth clime as the central and most noble part of the world may have far-reaching implications. Muslim authors tried to reconcile this particular vindication of Iranian culture with their religious beliefs by resorting to a double omphalic system in which prophecy occupies the position of superiority. Thus it becomes possible to harmonize the centrality

1 On the Iraqi school, see Ahmad, 'D̲j̲ug̲h̲rāfiyā', in *EI2*, vol. 2, pp. 579–81.

2 Cf. below pp. 69–70; and ʿAlī b. al-Ḥusayn al-Masʿūdī, *Kitāb al-Tanbīh wa'l-ishrāf*, p. 31, where the author claims that the fourth clime, called 'Bābil', is the most excellent.

of both Mecca and Baghdad. In his system, Ibn Khuraddādhbih went so far as to use geographical features, because he claimed that the 'belt of the Earth' (*minṭaqat al-arḍ*) stretches from Mecca to Baghdad, connecting both omphalic metropolises.[1]

But, despite the emphasis on the privileged status of the fourth clime, no reference is made to an omphalic metropolis in the Ikhwān's epistle on geography. In Epistle 22: 'On the Modalities of the Coming-to-Be of the Animals and of their Kinds', the Ikhwān state that the creation of the first earthly Adam took place on Mount Yāqūt, located to the south of the Equator, and which de Callatäy has identified as Sri Lanka, considering it to be a reference to the 'axis of the World'.[2] Both this location and the concept of the 'axis of the world' may have been borrowed from Indian geographical traditions, but seem not to have been taken into consideration when writing this epistle on geography.

In keeping with the tradition according to which the fourth clime is the best, the Ikhwān do not make any attempt to justify their description of the fourth clime as the most excellent. In view of the geographical determinism with which pre-modern ethnography was imbued, they explicitly relate the equilibrated climatic conditions of this region to the balanced nature of their inhabitants, and emphasize the excellence of the fourth clime also in terms of the superior virtue of its people. The encomium of the inhabitants of this region can also be found in other authors, such al-Yaʿqūbī, who praises the people of Baghdad in similar terms.[3] But the Ikhwān go a step further. In a striking statement, which has no parallel in cognate geographical works or, for that matter, in any other Muslim source, the authors of the epistle 'On Geography' claim that the fourth clime is the clime of the prophets (*anbiyāʾ*).[4] This statement is in keeping generally with how the fourth

1 ʿUbayd Allāh ibn ʿAbd Allāh ibn Khurradādhbih, *Kitāb al-Masālik waʾl-mamālik*, ed. M. J. de Goeje. *Bibliotheca Geographorum Arabicorum*, vi (Leiden: Brill, 1889), p. 67; Montgomery, 'Serendipity, Resistance and Multivalency', p. 190 f.

2 Cf. G. de Callataÿ, *Ikhwān al-Ṣafāʾ: A Brotherhood of Idealists on the Fringe of Orthodox Islam* (Oxford: Oneworld, 2005), p. 28.

3 Cf. Below, p. 69; and Aḥmad ibn Abī Yaʿqūb al-Yaʿqūbī, *Kitāb al-Buldān*, ed. W. T. Juynboll (Leiden: Brill, 1861), p. 233.

4 Baffioni has compared the geography of the Ikhwān with al-Muqaddasī, but she does not offer any interpretation of this reference to the prophets; see C. Baffioni, 'Il "quarto clima" nell'Epistola sulla Geographia degli Iḫwān al-Ṣafā'',

clime is treated by the Ikhwān in other epistles, but it is puzzling. How can it be reconciled with the location of the sacred city of Islam? In all extant versions of this epistle, Mecca belongs to the second clime and, consequently, and by implication, Muhammad would be excluded from the land of the prophets. By linking geographical position and prophecy, the Ikhwān seem to preclude any possibility of harmonizing ideology and religion by means of a double omphalic system such as that of Ibn Khurradādhbih. But they do not make any attempt to explain this oddity either.

On the basis of the possible identities of the members of the group and the vocabulary used in the epistles a Persian origin has plausibly been suggested for the Ikhwān. Perhaps this is why the fourth clime is depicted as it is? It would certainly introduce a prominent ideological statement that clearly needs further study. Prophets certainly play an important role in the Shuʿūbiyya polemics of the third/ninth and fourth/tenth centuries when comparing the virtues of Arab and non-Arab peoples, since there are only four Arab prophets — Hūd, Ṣāliḥ, Ismāʿīl, and Muhammad — a small number in comparison with those of non-Arab origin.[1] Moreover, there are numerous attempts in this period to link the Persians to Abraham by laying claim to a descent from Isaac.[2] We cannot completely discard the possibility that this nativist Persian discourse might have influenced the depiction of the fourth clime in the epistle 'On Geography', but the overall treatment of prophecy here seems to point in a different direction.[3]

In fact, the depiction of the fourth clime as 'the clime of the prophets' does not seem to occur in any of the works vindicating the centrality of Iran. I am only aware of two other works which depict the fourth clime in these terms. Both are late sources that have striking parallels

---

in *Oriente Occidente. Scritti in memoria di Vittorina Langella*, ed. F. Bencardino (Naples: Instituto Universitario Orientale, 1993), pp. 45–60.

1 See, for instance, ʿAbd Allāh ibn Muslim ibn Qutayba, *Faḍl al-ʿarab waʾl-tanbīh ʿalā ʿulūmihā*, ed. M. Khāliṣ (Abu Dhabi: Al-Majmaʿ al-Thaqāfī, 1998), p. 51.

2 Ibid., p. 47. On this claim, see S. Savant, 'Isaac as the Persian's Ishmael', *Comparative Islamic Studies*, vol. 2, 1 (2006), pp. 5–25.

3 On the Iranian nativist discourse, see P. Crone, *Nativist Prophets of Early Islamic Iran* (Cambridge: Cambridge University Press, 2012).

with the epistle 'On Geography' and suggest a relationship of some sort, be it direct or indirect.

The earliest work, written at the beginning of the seventh/thirteenth century, is Ibn Shaddād's *al-Aʿlāq al-khaṭīra fī dhikr umarāʾ al-Shām waʾl-Jazīra*, a history of Syria and Mesopotamia. Ibn Shaddād begins his description of Aleppo by commenting on the characteristics of the fourth clime. As authorities he cites Hermes, to whom he attributes the opinion that the fourth clime is the most excellent, and Ptolemy, from whom he allegedly quotes some passages which have clear similarities with the text of the Ikhwān al-Ṣafāʾ, such as the determination of the middle point of the clime — 36° in Ibn Shaddād's text, 36° 50' in the Epistle on Geography — and the number of cities contained in it — two hundred and twelve in both texts; the number of mountains given in both works, however, differs. The passage concerning the prophets quoted in the *Aʿlāq al-khaṭīra* has the same wording that the one in 'On Geography': 'This clime is the clime of the prophets (*anbiyāʾ*) and the wise men, because it is the central [clime], with three climes to the South and three to the North and because it is also in the division of the most brilliant body (*al-nayyir al-aʿẓam*).'[1]

The second source that contains a reference to the prophets is also a history of Aleppo written in the seventh/thirteen century: Ibn al-ʿAdīm's *Bughyat al-ṭalab fī taʾrīkh Ḥalab*. In this case, the information about the fourth clime comes from a treatise entitled *Risāla fī dhikr al-dunyā wa mā fīhā min al-aqālīm waʾl-jibāl waʾl-anhār waʾl-bilād* (*Epistle on the Earth and all the climes, mountains, rivers and lands it contains*), the identity of the author being unknown to Ibn al-ʿAdīm. The passages are from the source used by Ibn al-Shaddād, but Ibn al-ʿAdīm's quote is longer, allowing us to discern further similarities with 'On Geography'. The text quoted in the *Bughyat al-ṭalab* cites the same authorities: there is a reference to Hermes as the first author to use the division of the world into seven climes, and the authority of Ptolemy is invoked in the discussion of the division of the inhabited

1 Bahāʾ al-Dīn Yūsuf ibn Rāfī ibn Shaddād, *al-Aʿlāq al-khaṭīra fī dhikr umarāʾ al-Shām waʾl-Jazīra*, ed. Y. Z. ʿAbbāra (Damascus: Manshūrāt Wizārat al-Thaqāfa, 1991), p. 36.

quarter into seven climes and the characteristics of the fourth clime, to which Aleppo belongs.

The textual correspondences with 'On Geography' are to a great extent literal. The discussion of the division of the climes and of the concepts of latitude and longitude corresponds, almost *verbatim*, to the equivalent section in the epistle.[1] The description of the limits of the fourth clime in Ibn al-ʿAdīm's text, however, does not correspond to the wording of 'On Geography' nor are the latitude co-ordinates similar — Ibn al-ʿAdīm states that the longitude of the middle point in this clime is 36°, whilst it is 36° 50' according to the Ikhwān; the length of this clime is 5° 4' 30" in Ibn al-ʿAdīm, whilst in 'On Geography' this value is 5° 30'.[2] The following section, in which the lands in the clime are listed, largely corresponds — though not *verbatim* — to the text of the Ikhwān, including the unexpected mention of the city of al-Qādisiyya in the Maghrib.[3] The number of cities given in both texts is the same — two hundred and twelve — and although the number of mountains in Ibn al-ʿAdīm's text does not correspond to the number given in 'On Geography', a corruption of the text can easily be perceived.[4] After the geographical description of the fourth clime, this division is described as the clime of the prophets using the same rendition that we find in Ibn Shaddād's work and the epistle 'On Geography'.[5]

We cannot but speculate on the origin and the authorship of the source consulted by Ibn Shaddād and Ibn al-ʿAdīm. On the one hand,

1 Kamāl al-Dīn ʿUmar ibn Aḥmad ibn al-ʿAdīm, *Bughyat al-ṭalab fī taʾrīkh Ḥalab*, ed. S. Zakkār (Beirut: Dār al-Fikr, 1988), vol. 1, p. 443, lines 12–15, which correspond to 'Epistle on Geography', pp. 57–62 below, with exception of the lengths of the first and the seventh clime, omitted in Ibn al-ʿAdīm's text.

2 Respectively Ibn al-ʿAdīm, *Bughyat al-ṭalab*, vol. 1, p. 444, lines 1–4, which correspond to the epistle 'On Geography', p. 69 below.

3 See below p. 69.

4 The references to the mountains and the rivers are amalgamated and the number of mountains has been omitted and subsequently confused with the number of rivers, which is given later; thus the number of mountains in Ibn al-ʿAdīm's – and probably also in Ibn Shaddād's – text refers, in fact, to the number of rivers. This number (twenty-two) is similar to that given in 'On Geography'. The editor of Ibn al-ʿAdīm has tried to solve this textual oddity by reading *yahr* (width) instead of *nahr* (river), cf. Ibn al-ʿAdīm, *Bughyat al-ṭalab*, vol. 1, p. 444, n. 1.

5 Ibn al-ʿAdīm, *Bughyat al-ṭalab*, vol. 1, p. 443 f.

the reference to the prophets, the other textual parallels, and Ibn al-ʿAdīm's reference to the anonymity of his source seem to point to the epistle 'On Geography' or a recension of this work. On the other hand, the discrepancies concerning the co-ordinates and the reference to Hermes might suggest that these authors used another source.[1] If this were the case, this could have been either a text that used the epistle 'On Geography', or a text written prior to the composition of the works of the Ikhwān al-Ṣafāʾ, in which case the depiction of the fourth clime as the clime of the prophets would not be original to the Ikhwān. Nevertheless, the presence, direct or indirect, of a version of the Ikhwān's epistle 'On Geography' in these texts on Aleppo seems the most plausible explanation for the textual correspondences.

The Ikhwān's reference to the geographical origin of the prophets may be partly explained if we consider the strong geographical and cosmological determinism that underpins all the epistles. In the last section of the epistle 'On Geography', the revelation and the advent of prophets are also subordinated to the influence of the felicitous stars; prophets are no different, in this regard, from other worldly phenomena.[2]

Moreover, the Ikhwān's gnostic elitism is a further characteristic that should be considered when analysing their treatment of prophecy and prophets. In the epistle 'On Geography', we might be tempted to read their portrayal of the knowledge of the signs of God's Creation (*āyāt*) as endowed with a superior value that might relegate the prophets to an ancillary position. Though when we place the value of such observation alongside iterations of the same notion in other epistles by the Ikhwān we come to see that the men of wisdom have a share of prophecy and it is thus that they are empowered to read the signs correctly. For the

1 The reference to Hermes is, nonetheless, not unusual in early Islamic astronomical and geographical texts, cf. K. van Bladel, *The Arabic Hermes* (Oxford: OUP, 2009), p. 29 f. The Ikhwān discuss the astrological knowledge of the ancients – including Hermes – in the epistle 'On Magic', cf. *Rasāʾil Ikhwān al-Ṣafāʾ*, ed. Buṭrus Bustānī, 4 vols. (Beirut: Dār Ṣādir, 1377/1957), vol. 4, p. 295ff. See also Godefroid de Callataÿ and Bruno Halflants, *On Magic. An Arabic Critical Edition and English Translation of Epistle 52a* (Oxford: OUP–IIS, 2011), p. 116 ff.

2 See below, p. 75.

Ikhwān salvation is potentially available to all who study the noble sciences and apply them to the soul through action and thought.

The allegory of the king and his wondrous city is a vindication of the noble sciences needed to understand the semiotics of God's Creation.[1] In this allegory, the king represents God and the visitors are the believers who, like the Brethren, engage in the critical examination of the world in order to understand the signs of God, represented in the allegory by the inventions and prodigies created by the king. The Ikhwān refer to the marvels built in the city as 'inventions which can only properly be described if you have seen them for yourself' after having made the effort to travel to the city. The antithesis between direct observation and report is a common *topos* in geographical writings, and it is possible that the phrase may be purely stylistic.[2] In the context of this epistle, however, it would not be unreasonable to read this statement as an exposition of two different kinds of revelation: the one conveyed by the signs of God's Creation that those initiated in the noble sciences can understand, and that conveyed to all people alike by the prophets whose speech is distinguished by the ability to appeal using the same words to people of widely differing capacities.

In order to combine these maximal notions of prophecy and wisdom the Ikhwān identify the fourth clime as the home of prophecy. They seem to take it as a given that this does not exclude the prophecy of the Prophet Muhammad. The reading of God's signs and prophetic revelation thus become equally valid.

1.4.5. The Semiotics of Creation: Geography in the Context of the Epistles

If we only consider the information we might think of as geography proper, we note that the epistle 'On Geography' is an epitome of the geographical chapters of al-Farghānī's *Jawāmiʿ*, to which several tables, taken from al-Khwārizmī's reformulation of Ptolemy's *Geography*, have been added. There is certainly no original contribution to mathematical

1 See below, pp. 59–60.
2 For a discussion of this antithesis, see T. Zadeh, *Mapping Frontiers Across Medieval Islam* (London: I.B. Tauris, 2011), p. 7ff.

geography in this epistle, but that was not the purpose of the authors. Nor does this treatise exhibit the generic characteristics of other contemporary geographical works written in a form closer to *adab*. It does not contain historical information, poetry or anecdotes, and, if we exclude the sections where the reader is admonished to follow the path of the Brethren by means of allegories, the presentation of the geographical data would correspond to what scholarship has regarded as purely technical exposition. This seems to have been the conclusion of the majority of the scholars who have studied this epistle, and also that of its German and Italian translators, who do not translate the sections they do not consider proper to the subject.

The limitations of categorizing these texts in terms of the dichotomy of technical and non-technical information become evident when we avoid the decontextualization of the geographical information and instead read the treatise as a coherent textual unit in the holistic framework of the *Rasāʾil Ikhwān al-Ṣafāʾ*. The authors conceive of geography as one of the propaedeutic sciences (*ʿilm al-riyāḍiyyāt*), as an introduction to astronomy, despite being relegated to the fourth place after numbers, geometry and astronomy.[1] It would be a mistake, however, to consider geography as a simple *ancilla astronomiae*. The introductory character of this discipline is determined not only by the technical skills that its study imparts, but also by the Neoplatonic notion of progress in the form of cosmic ascent with which all the writings of the Ikhwān are imbued.

For the Ikhwān al-Ṣafāʾ, geography is a prerequisite to the study of astronomy. The cosmic dualism that opposes the superior world of the spheres to the sub-lunar world of coming-to-be and passing-away has a direct epistemological correspondence with the place occupied by the sciences in the classification of the *Rasāʾil*. The subject of geography is the sub-lunar world, the temporary realm in which we live, and therefore it provides a kind of knowledge which is of inferior value but nonetheless necessary in order for man to advance towards knowledge of the celestial sphere. It does not mean that this knowledge is purely instrumental or empirical. On the one hand, geography is the discipline

1 On the order of the epistles, cf. A. Hamdani, 'The Arrangement of the *Rasāʾil Ikhwān al-Ṣafāʾ*', in *The Ikhwān al-Ṣafāʾ and their* Rasāʾil. *An Introduction*, ed. N. El-Bizri (Oxford: OUP–IIS, 2008), pp. 83–100.

that allows us to understand the harmonic correspondence between the temporary world and the eternal realm; it provides us with the fundamental concepts required to study astronomy because the Earth is a mirror of the celestial sphere. On the other hand, the knowledge of the world in which we live has a soteriological dimension, since it can also be a first step towards the realization of the divine in the rational order of the universe. The study of the phenomenal world is, in this respect, a path to salvation and, when seen from this perspective, the noble sciences vindicated by the Ikhwān al-Ṣafāʾ reveal themselves as a method by which to understand the semiotics of God's Creation.

The ethical dimensions of the study of geography are illustrated by the Ikhwān al-Ṣafāʾ by means of the aforementioned allegory of the wise king who builds a fantastic city where he displays all his creations.[1] Those who pass through the city with no other concern than to satisfy their basic appetites are like those who do not occupy themselves with understanding God's signs while in this world. The realization of God's wisdom in the perfect arrangement of Creation has evident ethical and religious implications and is deeply rooted in the Neoplatonic framework of the *Epistles*. As the Ikhwān al-Ṣafāʾ put it in the proemium to the treatise, scrutiny of the temporary world compels the souls of mankind to reflect upon the harmony of the celestial realm, thus elevating them from the sleep of negligence and the slumber of ignorance.[2]

In contrast to other contemporary geographical works, where God's wisdom seems to manifest itself in the incommensurability of His Creation and the infinite variety of the wonders catalogued by the geographers,[3] the Ikhwān's epistle 'On Geography' celebrates the order of the universe and the possibility of understanding God's plan by cultivating the noble sciences. In this regard, Ptolemaic mathematical geography provides the ideal vehicle for emphasizing

1 See below, pp. 59–60.
2 See below, pp. 49–50.
3 See, for instance, Muḥammad ibn Aḥmad Shams al-Dīn al-Muqaddasī, *Kitāb Aḥsan al-taqāsīm fī maʿrifat al-aqālīm* (Leiden: Brill, 1967), which ends each chapter with an account of wonders (*ʿajāʾib*). On the relation of this notion with geography see André Miquel, *La géographie humaine jusqu'au milieu du 11e siècle*, 4 vols. (Paris: Éditions de l'EHESS, 1973–1988), vol. 4, pp. 51–54 and 95–122.

both the rationality of Creation and the profound religious and ethical dimensions of this discipline. This conception of geography agrees with the Pythagorean philosophical and theological doctrines on which the Ikhwān relied in order to categorize the hierarchies of the world and the ranks in the chain of being. The epistle is also a vindication of their faith, fully coherent with their mission to purify Islam. As the Qurʾanic verses they quote assert, nothing has been created without a known purpose: the initiation into the sciences, such as geography, that allows Muslims to understand God's designs is not only a scientific endeavour, but also a demonstration of piety.

It is evident that, for the Ikhwān al-Ṣafāʾ, geography is a science imbued with religious meaning, but did it have any political significance? The epistle 'On Geography' also parts ways with cognate works as regards the circumstances of its composition. Unlike other geographical treatises written by prominent members of the ʿAbbāsid bureaucracy and commissioned by the caliphs, these epistles are surrounded by mystery; little is known for certain about their authors, and there is no evidence suggesting a possible patronage of the work. The chorographical discourse of the Ikhwān entails a celebration of God's Creation, but in this case it is by no means consubstantial with the celebration of the ruler. It is true that their apparent detachment from the political sphere in Iraq, the anonymity of the authorship of the epistles, and the gnostic elitism that underpins the entire work, are political statements in themselves; but the actual significance of these aspects is lost in the absence of a clearer historical context, and it is difficult to imagine how geography could convey a political message under these circumstances. Yet for all that, the epistle 'On Geography' contains one of the most — if not the most — relevant political statements to be found in the *Rasāʾil*.

The context in which this declaration occurs is the discussion of the cycles of time, which serves as a corollary to the Ikhwān's considerations on the ascendancy of the stars over the Earth.[1] The alternate temporal cycles governed by fortunate and unfortunate stars influence the succession of opposite and inversely proportional worldly phenomena. Just as time has its cycles of day and night, and summer and winter, opposite

1 See below, pp. 76–77.

elements simultaneously increase and decrease in the same proportion: whenever one of them reaches its apogee, the opposite element reaches its nadir. This dynamic also explains political cycles. In one of the few deictic statements of the entire collection, the Ikhwān al-Ṣafā᾽ write that they are witnessing the turn of the evil-doers (*dawlat ahl al-sharr*) and that, by virtue of this cyclic law, once their apogee has been reached, their dominion will begin to decline in the same proportion as the power and influence of the opposite group, the virtuous (*ahl al-khayr*), will grow. The Ikhwān identify themselves with the *ahl al-khayr* and believe that, at the inception, the virtuous are little more than a small coterie of scholars. The identity of the evil-doers should be looked for, therefore, in the ruling dynasty at the time of the composition of the epistles.

In addition to this, we should also draw attention to the possible Iranian ideology underpinning the work. As has been referred to above, the consideration of the fourth clime, which contains the Sawād and Baghdad, as the most excellent is a recurrent *topos* in the writings of Iranian geographers with clear ideological overtones, and the depiction of this clime as the clime of the prophets could have also been partly influenced by Iranian ideology. The passages describing the advent of the *ahl al-khayr* and the decline of the *ahl al-sharr* also have clear dualistic overtones. It is possible to find dualistic parallels in Neoplatonic doctrine, but in the context in which this discussion appears — amid references to night and day, and light and darkness — a possible connection with Iranian sources may not have passed contemporary Muslim readers unnoticed.[1]

1 On the Iranian tradition and its relation to cosmography, see S. H. Nasr, *The Islamic Intellectual Tradition in Persia* (Richmond: Curzon Press, 1996), pp. 10–27.

(fol. 35a) [p. 158]

Epistle 4

On Geography

On the Shape of the Earth and the Climes

(Being the fourth epistle in the first section
of the Propaedeutic Sciences)

*In the name of God the Compassionate, the Merciful**

The procedure of our Brethren, may God aid them and us with a spirit of His, is the examination of all existents, the investigation of their first principles, the cause of their existence, and the ranks of their order, and the explanation of how their effects and their causes are connected, by the permission of their Creator, sublime is His praise! Therefore we need to discuss the condition of the Earth, how it is shaped, and the reason for its stationary position [p. 159] in the centre of the universe.

Knowledge of the Earth and of how it is stationary in the air belongs to the noble sciences, because the Earth is what our bodies stand on, it is where our bodies start their existence and grow, and whence they derive the matter for their continuance; and it is thither that they return when they are separated from our souls. Furthermore the study of this science is a means for raising the preoccupations of our souls to the world of the celestial spheres, the abode of the exalted believers.[1] Moreover (fol. 35b) when our thoughts roam through the plain of the spiritual beings, and we make frequent reference to the world (by which I mean the world of the celestial spheres) our souls are given an opportunity to awaken from the sleep of negligence and the slumber of ignorance. This invites them to rise from the world of coming-to-be and passing away[2] to the world of continuance and duration, and makes them desire the journey from the world of the

* The title in the Atif MS runs as follows: 'The Epistle "On Geography", i.e., the shape of the Earth and the climes, the fourth in the First Section of the Propaedeutic Disciplines' and is followed by the eulogy: *In the name of God the Compassionate, the Merciful.*

1 The *'illiyya* is the place in the Seventh Heaven to where the souls of the believers ascend; the *'illiyyūn* those who dwell there.

2 The world of coming-to-be and passing away is the earthly world, which is the sub-lunar sphere.

material bodies and the vicinity of the devils, to the world of the spirits and the vicinity of the angels who are close [to God].

In this epistle, we have discussed a number of points about how the Earth is shaped, and have described the inhabited quarter, with the climes, seas, mountains, land masses, rivers and cities that it contains, so that it can serve as a path for those beginning their study of the science of astronomy, of the arrangement of the celestial spheres, of the ascension of the zodiacal signs, and the revolution of the stars; to make it easier for those studying them to conceive of them; and so that it would be easier for those who ponder upon *the dominion of the heaven and of the earth*[1] to conceptualize them, those who say: *Our Lord, Thou hast not created this for vanity. Glory be to Thee! Guard us against the chastisement of the Fire* [p. 160].[2]

Description of the climes, of the inhabited quarter of the Earth with its mountains, seas, land masses, rivers, cities, and of the seas with their islands and cities

Firstly, before our description, we need to discuss the characteristics of the Earth, its six cardinal points and how it is fixed in the air. The six cardinal points are: east, west, south, north, up, and down. The east is where the Sun rises, and the west where it sets; the south is where Canopus revolves, and the north where the Polar Star and the *Farqadān*[3] [revolve]; up is the region of the sky; and down is the region of the centre of the Earth.

The Earth is a circular body, like a sphere. It is stationary in the air, by the permission of God the Exalted, with all its mountains, its

1 Reference to Qur'an 7:185. For the quotations from the Qur'an we have used Arberry's translation: A. J. Arberry, *The Koran Interpreted* (London: Allen & Unwin, 1955).

2 Qur'an 3:191.

3 The *Farqadān* — lit. 'the two calves' — are the two brightest stars in Ursa Minor, namely, η Ursae Minoris and ζ Ursae Minoris. See Fuat Sezgin, *Mathematical Geography and Cartography in Islam and Their Continuation in the Occident* (Frankfurt: Institute for the History of Arabic-Islamic Science, 2007), vol. 2, p. 186.

seas and landmasses, its inhabited and uninhabited areas. The air surrounds the Earth in all directions: east, west, south, and north, on this side, and the other side.[1]

The distance of the Earth from the sky is equal in all directions. The circumference of the surface of the Earth at its greatest is 20,400 miles,[2] which is 6,800 *farsakhs*.[3] The diameter of this circumference, the diameter of the Earth, is 6,500 miles,[4] approximately 2,170 *farsakhs*. The centre [of the Earth] is an imaginary point deep inside it, at the midpoint of its diameter. Its distance from the visible part of the surface of the Earth and the sea is equal in all directions. The lowest point is not constituted by any part of the visible surface of the Earth in any direction, as many of those who have no training in the discipline of geometry imagine. They imagine and hypothesize that the lowest part of the Earth is the surface of the Earth on the side opposite our location, that the air that surrounds that side is also lower than the Earth, and that the half of the sphere of the Moon that surrounds the air is lower than the air. They applied this to all the levels of the celestial spheres [p. 161], each lower than the other. So the consequence of this misconception is that the lowest point of all is that half of the all-encompassing celestial sphere which has always been the highest of the high.[5]

1 i.e., up and down, completing the six cardinal directions.

2 Other manuscripts give 30,400 miles [ج, خ], 24,000 miles [ع], and 20,431 miles [ن], with the Arab mile being approximately 1.6 km. Despite the discrepancies as regards the length of the circumference, most manuscripts give a measure of the diameter which is consistent with a value of 20,400, which accords with the estimation of al-Farghānī; see *Elementa Astronomica*, p. 30 (Arabic text). The actual length of the equatorial circumference of the Earth is 24,873 miles (40,030 km).

3 On the *farsakh*, or parasang, see N. El-Bizri (ed.), *On Arithmetic and Geometry; An Arabic Edition and English Translation of Epistles 1 and 2* (Oxford: OUP–IIS, 2012), p. 132, n. 41.

4 Other manuscripts give 6,501 miles [ز, ط, غ, ك] and 7,363 ,[ل] and 4/11 (*sic*) miles [ع] and 6,401 miles [ن]. With the exception of manuscripts [ع] and [ن], these figures approximate to a circumference of 20,400 miles. The actual diameter of the Earth at the equator is approximately 7,926 miles (12,756 km).

5 The all-encompassing celestial sphere (*muḥīṭ*) is the outermost of the celestial spheres. According to the Ikhwān's ontological doctrines, it was brought into existence by the Universal Soul out of love for God. The 'highest of the high' and the 'lowest of the low' are two central concepts in the cosmological model of

The situation is not as they imagine it. This is a weak idea given credence by men from their childhood without reflection or solid proof. When someone has had instruction in the science of astronomy, it becomes evident that the situation is the opposite of what he had previously imagined. In truth, the lowest [place] on Earth, from any part of the Earth, is in the depths of its interior next to its centre, because the centre of the Earth is the lowest point of all. 'Up' is the visible surface in direct contact with the air in all directions. The surrounding air is above the Earth in all directions, the sphere of the Moon is above the air, the sphere of Mercury is above the sphere of the Moon and, according to this schema, the rest of the spheres are above one another until the ninth celestial sphere which is above all the other spheres, and is the highest of the high. The centre of the Earth, the lowest point of all, is opposite it.

You should know, dear Brother, that a person standing at any point on the surface of the Earth, be it east, west, south, or north, on this side or the other side, will always be standing upon the Earth, his head up towards the sky, and his feet down towards the centre of the Earth. He will see half the sky. The other half will be concealed from him by the curvature of the Earth. If he moves from one place to another, the part of the sky that becomes visible to him is in proportion to the amount of the other side concealed from him. This proportion is 19 *farsakh*s to 1 degree [p. 162].

The reason for the stationary position of the Earth in the middle of the air

There are four opinions as to why the Earth is stationary in the middle of the air. Some claim that it is stationary because the celestial sphere pulls on it from all directions with equal [force], so it has necessarily to be stationary in the middle by virtue of the equal force of the pull from all sides. Some claim that [the celestial sphere] pushes the Earth in a similar manner so that it has to remain stationary by virtue of the

the Ikhwān; see Seyyed Hossein Nasr, *An Introduction to Islamic Cosmological Doctrines* (New York: SUNY Press, 1993), p. 70.

equal force of the push on all sides. Others claim that it is stationary in the middle of the air because the centre [of the Earth] pulls all its constituent parts from all sides towards the middle, since the centre of the Earth is also the centre of the sphere and is a magnet for the heavy bodies.[1] This means that all parts of the Earth are pulled to the centre of the Earth because they are heavy. One part precedes the others and when it reaches the centre it stops. The remaining parts come to a stop around it, i.e., around the [central] point, but as each part tries to reach the centre, the Earth and all its parts become a single sphere. Since the constituent parts of water are lighter than the parts of earth, water comes to a stop on top of earth; and, since the parts of air are lighter than the parts of water, air comes to a stop on top of water; and since the parts of fire are lighter than the parts of air, fire ascends to the region of the sphere of the Moon.

The fourth reason adduced to explain the stationary position of the Earth in the middle of the air is the special characteristic of the place especially appropriate for it (fol. 36a). The Blessed and Exalted Creator assigned to each of the universal bodies, i.e., fire, air, water, and earth, a specific place which is the place especially appropriate to it. He assigned a specific place to the Moon, Mercury, Venus, the Sun, Mars, Jupiter, and Saturn, each in its celestial sphere, where they are fixed whilst the sphere causes them to revolve as it revolves. This account is closest to the truth, because this cause continues throughout the arrangement of the nine celestial spheres, the fixed and the movable stars, and the four basic elements — I mean, fire, air, water, and earth. For Blessed and Exalted God, has assigned each and every existent a place more specific to it than any other place [p. 163], and a known rank, more appropriate to it than any other rank.[2]

1 The Ikhwān use the simile of a magnet to discuss the stationary position of the Earth by applying the Aristotelian explanation of gravity, the so-called theory of the 'natural place'. Magnetism is discussed on another occasion in the collection of epistles: see *Rasāʾil Ikhwān al-Ṣafāʾ*, ed. Buṭrus Bustānī, 4 vols. (Beirut: Dār Ṣādir, 1377/1957), Epistle 19, vol. 2, p. 110ff. Al-Muqaddasī also refers to the centre of the Earth as a magnet in his geographical treatise: *Aḥsan al-taqāsīm fī maʿrifat al-aqālīm*, vol. 5 of *Bibliotheca Geographorum Arabicorum*, ed. M. J. de Goeje, 8 vols. (Leiden: Brill, 1967), p. 58.

2 This use of the participle 'known' has Qur'anic overtones. It seems to echo the phrase *maqām maʿlūm* ['known station'] in Qur'an 37:164–66, the verse added

The Earth and its division into quarters

Half the surface of the Earth is covered by the waters of the Great Encircling Sea[1] and the other half is uncovered, like an egg, half of which is submerged, with the other half protruding from the water. Half of this uncovered half is uninhabited, and lies to the south of the equatorial line. The other half, the inhabited quarter, lies to the north of the equatorial line. The equatorial line is an imaginary line that runs from east to west where the Head of Aries[2] revolves. Night and day are always equal at this line, and the two poles there are fixed to both horizons: one is where Canopus revolves in the south, the other is in the north where Capricorn rises.

Here is the diagram:

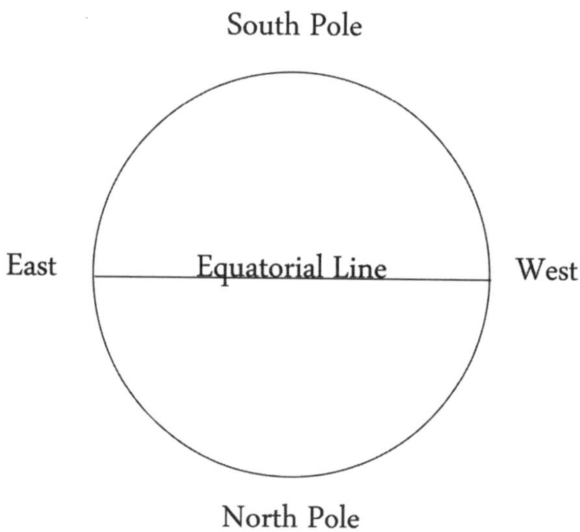

in manuscript [ج]: *He, may His praise be glorified, says: 'None of us is there, but has a known station. We are the rangers, we are they that give glory'* [i.e. the angels].

1 *al-baḥr al-muḥīṭ* (the Encircling Sea), also known as *al-baḥr al-akhḍar* (the Dark Sea) or *Uqyānūs* (the Ocean), was thought to enclose the habitable world; see D. M. Dunlop, 'al-Baḥr al-Muḥīṭ', in *EI2*, vol. 1, p. 934.

2 The term *ra's burj al-ḥamal* (the Head of Aries) refers to the three stars forming the two horns of the constellation of Aries, which in Arabic bear the names of *al-ḥamal* and *al-sharaṭān*. See P. Kunitzsch, *Untersuchungen zur Sternnomenklatur der Araber* (Wiesbaden: Otto Harrassowitz, 1961), p. 93.

Description of the inhabited quarter of the Earth

There are seven great seas in the northern, inhabited quarter. Each sea has a number of islands. The size of each island is from 20 to 100 or 1,000 *farsakhs*. [The seas] are: the Sea of the Romans,[1] which contains around 50 islands [p. 164]; the Sea of the Slavs,[2] which contains 30 islands; the Sea of Jurjān,[3] which contains five islands; the Sea of al-Qulzum,[4] which contains 15 islands; the Persian Sea, which has seven islands; the Indian Sea,[5] which has approximately 1,000 islands; and the Sea of China, which contains around 200 islands.

In this quarter there are also more than 15 small salt- and fresh-water seas, from 20 to 100 or 1,000 *farsakhs* in size. The Western Sea,[6] the Sea of Gog and Magog,[7] the Sea of the Zanj,[8] the Sea of Zābaj,[9] the

1 The name used here to designate the Mediterranean is *baḥr al-rūm* (the Sea of the Romans), on this name see Dunlop, 'Baḥr al-Rūm', in *EI2*, vol. 1, pp. 934–36. The Byzantines were known as the Romans.

2 This is presumably the Baltic Sea.

3 The Aral Sea.

4 The Red Sea; see C. H. Becker, 'Baḥr al-Ḳulzum', in *EI2*, vol. 1, pp. 931–33.

5 Lit., 'the Sea of Sind and Hind', on this name see Hartmann and Dunlop, 'Baḥr al-Hind', in *EI2*, vol. 1, pp. 930–31.

6 Lit., 'the Sea where the Sun sets', i.e. the Atlantic Ocean.

7 Gog and Magog are the names of apocalyptic peoples that occur in Qur'anic eschatological accounts concerning Alexander the Great, who built a barrier against them (Qur'an 17:93–98). The location of these tribes is unclear. The Muslim geographers located them either in the furthest north, beyond the lands of the Slavs, or in the eastern part of the land of the Turks, adjacent to China. In this epistle they are listed together with the Chinese as peoples of the seventh clime [see below p. 74], but the location of the Sea of Gog and Magog remains unclear. For a survey of Arabic sources on these peoples, see E. J. Van Donzel and A. B. Schmidt (ed.), *Gog and Magog in Early Christian and Islamic Sources* (Leiden: Brill, 2009).

8 The term 'Sea of the Zanj' refers to the western part of the Indian Ocean. The name derives from the Land of the Zanj (Zanzibar); see Dunlop, 'Baḥr al-Zandj', in *EI2*, vol. 1, pp. 937–38.

9 Although the majority of the manuscripts used in this edition do not have a correct rendition of this toponym, it seems clear that it refers to the island of Zābij, or Zābaj. Situated in the Western Indonesian Archipelago, it is mentioned in early Arabic geographical works, but its identification remains unclear. Some authors have suggested that it was the Arabic name used to refer to Java, but it is also possible that it referred to Sumatra or Borneo, or to the whole of the Sunda archipelago; see G. R. Tibbetts and Sh. M. Toorawa, 'Zābadj', in *EI2*, vol. 11, pp.

Dark Sea,[1] and the Encircling Sea lie outside this inhabited quarter. These salt-water seas are branches and tributaries of the Encircling Sea.

In this quarter there are also 200 large mountain ranges. Some extend from 20 to 100 or 1,000 *farsakh*s. They are of different kinds. Some stretch from east to west, or from south to north. Some stretch obliquely between east and south, some between east and north. Some cross the settled lands, cities, and villages; some cross the dry, arid plains; some cross islands and the seas.

This quarter has around 240 rivers, from 20 to 100 or 1,000 *farsakh*s in length. Some flow from east to west, some from west to east, some from south to north, some from north to south, and some flow obliquely between these directions. All these rivers begin in the mountains and flow into the seas, torrent beds, or lakes. In their course they provide water to cities, villages, and cultivated areas. The surplus water continues on to the sea, where it is mixed with sea-water. Then it turns into mist, which ascends into the air, and clouds are formed. The winds drive the clouds to the mountain-tops and the plains, where it rains, watering the land and filling the valleys and rivers. Then the water returns [p. 165] to the sea again. Such is the arrangement of the Almighty, the Omniscient.

In this quarter there are seven climes which incorporate approximately 17,000 large cities ruled by approximately 1,000 kings. All this is contained in one quarter of the surface of the Earth. This characteristic, however, does not apply to the other three quarters.

367–39. The mediaeval translation of this epistle renders this name as 'mare Ranig'; see P. Gaultier Dalché, 'Epistola Fratum Sincerorum in Cosmographia', *Revue d'histoire des texts*, vol. 18 (1988), p. 156.

1 In Arabic geographical sources the name 'Dark Sea' is often used to refer to the Encircling Ocean [*al-muḥīṭ*], which is the next sea listed here. In this context, the term 'Dark Sea' refers to the African littoral of the Indian Ocean; see C. A. Nallino, 'Al-Ḫwārizmī e il suo rifarcimento della Geografia di Tolomeo', in *Atti della Real Accademia dei Lincei*, anno 291, serie V, *Memorie delle Classe di Scienze Morali* (1894), vol. 2, p. 28.

Description of the Seven Climes

The climes are seven divisions that have been drawn in the inhabited quarter of the Earth. Each clime is like a rolled out carpet; its longitude goes from east to west and its latitude goes from south to north. They differ in longitude and latitude. The first clime is the longest in longitude and latitude; from east to west it is around 3,000 *farsakh*s long, and, from south to north, around 150 *farsakh*s wide. The seventh clime is the shortest in longitude and latitude; from east to west it is approximately 1,500 *farsakh*s long, and, from south to north, approximately 70 *farsakh*s wide. The differences in longitude and latitude between the other climes are represented as follows:

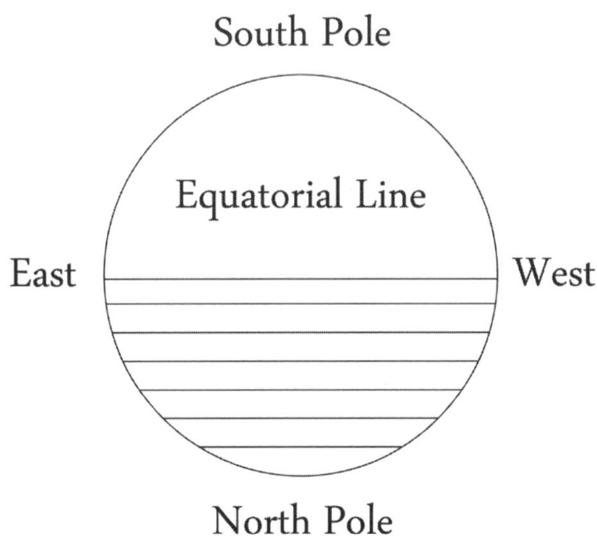

(fol. 36b) [p. 166]

Dear, esteemed Brother, may God aid you and us with a spirit of His, may it please you to know that these seven climes are not natural divisions, but imaginary lines drawn by the ancient kings who traversed the inhabited quarter of the Earth, kings such as Alexander the Greek,

Tubbaʿ al-Ḥimyarī,[1] Sulayman b. Dāwūd the Israelite, Afrīdhūn the Nabatean,[2] Ardāshīr ibn Bābakān the Persian, and others. [They drew them] to mark the frontiers, the highways, and the kingdoms. They were prevented from travelling to the three remaining quarters by lofty mountains, unpassable roads, swollen seas, by extreme temperatures, hot and cold, and a darkness such as is found in the region of the north below the Pole Star, where the cold is extreme. For six months during winter it is perpetual night there. The air is pitch black, the water freezes from the cold, and plants and animals are destroyed. Opposite this region, in the south under Canopus, it is perpetual day for six months during summer. The air warms up and becomes a raging fire that scorches plants and animals with its heat. It is impossible to live or to travel there. In the western region, the Encircling Ocean blocks any further passage with its storm-tossed waves and profound darkness. In the eastern region, mighty mountains block the way. If you consider this, you will realize that people are restricted to the inhabited quarter of the Earth, and they know nothing about the remaining three quarters.

You should know that when compared to the celestial spheres, the Earth, even with all its mountains and seas, is no more than a point inside a circle.[3] There are 1,029 stars in the Celestial Sphere. The smallest star is 18 times larger than the Earth, and the biggest star 107 times. Because the Celestial Sphere is so vast and far away, the stars appear to our eyes like pearls scattered on a dark carpet. If a person reflects upon all this magnitude and the wisdom, majesty and greatness of the Creator become clear [p. 167], then that person will

1 'Tubbaʿ' is the title and the name of a number of Ḥimyarī kings mentioned in Qurʾan 44:37, who ruled in Yemen until the Abyssinian invasion in the 4th century CE. The king referred to in this epistle is the most famous of the Ḥimyarī rulers, Tubbaʿ Asʿad Kāmil al-Ḥimyarī, who is supposed to have ruled between the fourth and the fifth centuries CE. According to Arabic sources, he was the first to cover the Kaʿba with a *kiswa*.

2 Afrīdhūn or Farīdhun, son of Abtiyān, was one of the early Persian kings. Many legendary deeds have been credited to him in the sources, among them the composition of the Khwārizmian astronomical tables; see H. Massé, 'Farīdūn', in *EI2*, vol. 2, p. 798.

3 This simile is inspired by Euclidean geometry; on the influence of Euclid on the *Epistles* of the Ikhwān, see N. El-Bizri, 'Epistolary Prolegomena: On Arithmetic and Geometry', in *The Ikhwān al-Ṣafāʾ and their Rasāʾil: An Introduction*, ed. N. El-Bizri (Oxford: OUP–IIS, 2008), pp. 180–213.

awaken from the sleep of ignorance and the slumber of negligence, and realize that these things have been created for a great purpose.[1]

Chapter

You must know, dear Brother, that some come into this world and live here for a long time, occupied with nothing but eating, drinking, and copulating, always in search of pleasure, bent on amassing wealth and chattels, on acquiring buildings and estates, and looking for personal pre-eminence. Such a man wishes to live here for eternity. He abandons the quest for knowledge and does not try to understand the true reality of things. He fails to exercise his soul, and does not prepare any provisions for the journey to the next abode. Then, when [his] life ends, when the next world and death's agony[2] draw near (I mean the separation of the body from the soul) he departs from this world in ignorance, without having understood its shape, without having reflected upon the signs [3] on the horizons, without having learned the lesson of the conditions of its existents, and without having scrutinized the sensorially perceptible things he has experienced in it.

This is like a group of people who enter the city of a wise, just, mighty, and merciful king. He has built it in his wisdom, and furnished every part of it with his wondrous inventions, inventions which can only properly be described if you have seen them for yourself. He has prepared a table with food for those who arrive, and with provisions for those who depart. Then he summons his servants into his presence so that he can bestow upon them his generosity. He commands them to travel to this city which is on their way, so that they can see and observe its contents, and reflect upon his wondrous inventions, and

1 This statement has Qur'anic overtones, and manuscripts [ك, ف, ع, ز], and [ن] quote the verse Qur'an 44:39: *We created them not save in truth; but most of them know it not.*

2 Reference to Qur'an 50:19: *And death's agony comes in truth; that is what thou wast shunning!*

3 The term *āya*, pl. *āyāt*, refers to a 'sign' or a 'wonder' that may lead to the realization or understanding of God; the 'signs' referred to here are the wonders of God's Creation. On this concept, see Rippin, 'Āya', in *EI3*.

learn the lesson of its marvellous forms and designs. He wants their souls to concentrate on all this and for them to become wise, good, and noble, upon seeing it. For then, when admitted into his presence, they will be worthy of his generosity.

But one night a group of people enter [the city] and spend the whole night eating, drinking, entertaining, and diverting themselves. Then they leave the city at daybreak without even knowing which gate they came in by nor which gate they left by, and without having seen any of the products [p. 168] of [the king's] wisdom or the marvels of his fashioning. Thus the only benefit they obtained was the food and drink they enjoyed that one night and no more. This is the determining characteristic of the 'sons of the world' who enter it in ignorance, abide there in perplexity, and depart unwillingly, in complete denial of the next abode. God, sublime be His praise, says: *And whosoever is blind in this world shall be blind in the world to come, and he shall be even further astray from the way.*[1] And He admonished them with these words: *Deaf, dumb, blind; so they do not understand*[2] anything about the hereafter.

I urge you not to be one of them, dear, esteemed Brother. Be one of those whom God Almighty praises when He, sublime be His praise, says: *That is the Last Abode; We appoint it for those who desire not exorbitance in the earth, nor corruption. The issue ultimate is to the godfearing.*[3] Their words are quoted when the sons of this world[4] petition their Lord and say: *Would that we possessed the like of that Korah has been given! Surely he is a man of mighty fortune.*[5] But *those to whom knowledge had been given*[6] concerning the true state of the hereafter said: *'Woe upon you! The reward of God is better for him who believes, and works righteousness; and none shall receive it*[7] *except the steadfast'*; *none shall receive it, except a*

1 Qur'an 17:72.
2 Qur'an 2:171.
3 Qur'an 28:83.
4 Reference to Qur'an 8:67.
5 Qur'an 28:79. Korah [*Qārūn*] is a biblical figure who lived in the time of Moses. He is described in Q.28:76–82 as a man to whom God granted great treasure.
6 Reference to Qur'an 58:11.
7 Qur'an 28:80. The end of this verse is not given as part of this quotation; instead it is merged with the beginning of another Qur'anic verse: Qur'an 41:35.

man of mighty fortune.[1] May God guide you in the true direction, dear Brother, to the path of righteous correctness.

Since we have completed our discussion and description of the Earth and the inhabited quarter, we would like to discuss the seven climes and clarify their boundaries in terms of longitude and latitude, as well as the large cities, the mountains, and the long rivers that each contains.

Dear, esteemed Brother, may God aid you and us with a spirit of His, you should know that the boundaries of the seven climes are determined by the hours of the day and their intervals of increase. The explanation for this is as follows. When the Sun is at the beginning of the sign of Aries, the length of Day and Night and their hours are equal in all these climes; but when the Sun advances by a number of degrees in the signs of Aries, Taurus, and Gemini, the hours of the day vary in each clime, until the sun reaches the end of Gemini, which is the beginning of Cancer. Then the length of the day in the middle of the first clime is 13 hours, in the middle of the second clime, 13 and a half hours; in the middle of the third clime [p. 169], 14 hours; in the middle (fol. 37a) of the fourth clime, 14 and a half hours; in the middle of the fifth clime, 15 hours; in the middle of the sixth clime, 15 and a half hours; and in the middle of the seventh clime, 16 full hours. In those places where the longitude ranges from 66 degrees to as far as 90 degrees, it is always daytime. The explanation of why this is so is lengthy and is discussed in the *Almagest*.[2]

You should know that 'longitude' means the distance of every land and city from the Far West, and 'latitude' means the distance from the equatorial line. The equatorial line is where Night and Day are always equal. Latitude does not apply to cities on this line, and longitude does not apply to cities located in the Far West. There are 180 degrees from the Far West to the Far East. Each degree corresponds to 19 *farsakh*s. A city with a longitude of 90 degrees is situated on a midpoint between the east and the west. Cities with greater [longitudes] are closer to the east and those with inferior [longitudes] are closer to the west. Of any two cities, the city with a greater longitude and latitude is closer to the east and the north. As for the distance between them in terms of

1 Qur'an 41:35.

2 Ptolemy's *Almagest* saw several translations in the 3rd/9th century; see the Introduction to this volume, pp. 29, 31 above.

latitude, each degree corresponds to 19 *farsakhs* approximately. The distance between them in terms of longitude varies: for those cities located on the equatorial line, each degree of longitude corresponds to 19 *farsakhs*; for those in the first clime, each degree corresponds to 17 *farsakhs*; in the second, to 15 *farsakhs*; in the third, to 13 *farsakhs*; in the fourth, to 10 *farsakhs*; in the fifth, to 7 *farsakhs*; in the sixth, 5 *farsakhs*; and in the seventh, 3 *farsakhs*.

Names of large cities located in the climes: cities whose latitude is less than 12 degrees from the equatorial line, beginning in the East [p. 170]

Names of the cities	Longitude	Latitude
Kashmir in Hind		
The City of Perfumes in Hind		4° 30'
Hadhramawt in Yemen	71° 00'	14° 00'
Waghāra in Abyssinia		
Kūkū in Abyssinia		05° 00"
A city on the Western Sea		

Table 1

The First Clime

The first clime is governed by Saturn; its longitude from east to west is 9,000 miles[1] — 3,000 *farsakhs* — and its latitude from south to north is 440 miles — 142 *farsakhs*. Its first [longitudinal] boundary is at the equatorial line, where the elevation of the North Pole is 13 degrees less a quarter [12° 45'], and the longest day is 12 and three-quarter hours long [12h 45']. Its midpoint lies where the elevation of the pole over the horizon is 16 degrees and two-thirds [16° 40'], and the longest day is 13 hours long [13h 00']. Its second boundary lies where the elevation

1 Other manuscripts give 6,000 [ج] and 51,000 miles [خ].

of the pole is 20 and a half degrees [20° 30'], and the longest day is 13 and a quarter hours long [13h 15'].

There are around 20 mountain ranges in this clime, between 20, 100, and 1,000 *farsakh*s in extent. There are also about 30 long rivers, measuring between 20, 100, or 1,000 *farsakh*s, and there are around 50 large cities known of. This clime begins in the east, to the north of the Island of Rubies,[1] and stretches through China towards the south. Then it runs along the north of Sri Lanka,[2] crosses through the middle of Hind and the middle of Sind,[3] cuts through the Persian Sea towards the south of Oman, crosses the centre of Abyssinia, and there cuts across the Egyptian Nile [p. 171], and from there it stretches through Nubia, through the middle of the land of the Berbers and Libya,[4] and through the south of Mauritania, ending in the Western Sea. The majority of the people in these lands are black [in colouring].

1 The Ptolemaic term 'Island of Sapphires' was commonly rendered into Arabic as Island of Rubies [*jazīrat al-yāqūt*]. Al-Khwārizmī also uses the name 'Island of the Jewels' [*jazīrat al-jawhar*]; according to the map of the unicum manuscript of the *Kitāb Ṣūrat al-arḍ*, this island is located in the Encircling Sea, close to the equator with an approximate latitude of 2° and longitude of 173°; see Nallino, 'Al-Ḫwārizmī e il suo rifarcimento della Geografia di Tolomeo', p. 41; and Muḥammad ibn Mūsā al-Khwārizmī, *Kitāb Ṣūrat al-arḍ*, ed. H. von Mžik as *Das Kitāb ṣūrat al-arḍ des Abū Ǧaʿfar Muḥammad Ibn Mūsā al-Ḫuwārizmī* (Leipzig: Harrassowitz, 1926), Tafel 1.

2 In Arabic: 'the Land of Sarandīb'.

3 The Land of Sind refers to the region around the lower course of the Indus.

4 None of the manuscripts used in this edition has a correct rendition of the toponym; in some of the accompanying tables it occurs as *bilād lībiwānī* or *lībwānī*, as it does in the manuscript containing al-Khwārizmī's cartographical work; see von Mžik (ed.), *Das Kitāb ṣūrat al-arḍ*, p. 101.

List of names of cities in this clime, i.e., every city with a latitude of 13 [13°] to 20 degrees [20°],[1] beginning in the East

Names of the Cities	Longitude	Latitude
A city in furthermost China		
A city in the islands of China		
B-ā-y-w-q-ā [?] in Hind[2]	130° 00'	12° 00'
Aden in Yemen	65° 30'	13° 00'
Dongola in Nubia	53° 40'	14° 00'
The kingdom of Abyssinia		
The lands of the Berbers		

Table 2

The Second Clime

The second clime is governed by Jupiter. Its longitude from east to west is 9,600 miles[3] [p. 172], and its latitude from south to north 400

1 *Sic*. The second limit of the clime, as indicated in the description, is actually 20° 30'.

2 This may be Ayodhya, although there is significant divergence in the manuscripts as to how the name is written: *b-a-y-w-q-a* [أ], *?-a-?-w-f-a* [ج], *ḥ-a-y-?-w-r-a* [خ], *m-a-?-?-w-w-a* [ر], *?-a-?-?-w-q-a* [ط], *?-a-q-w-m-a* [ع], *b-a-n-b-w-f-a* [غ], *?-a-m-w-q-a* [ف], *?-a-?-?-w-f-a* [ك], *?-a-s-w-?-a* [ل]. Behind this divergence, may be the toponym *Kh-ā-n-f-w-ā* given by al-Khwārizmī in *Ṣūrat al-arḍ*, see von Mžik (ed.), *Das Kitāb ṣūrat al-arḍ*, p. 4.

3 Other manuscripts give: 1,600 miles [ج]; 8,200 miles [ن, ر]; 7,600 miles [خ, ز, ط, ك, ل]; and 9,686 miles [ع].

miles.[1] Its first boundary is the end of the clime of Saturn, where the elevation of the pole is (fol. 37b) 20 and a half degrees [20° 30'] and the longest day is 13 and a quarter hours long [13h 15']. Its midpoint is where the elevation of the pole is 24 degrees and 6 minutes [24° 6'] and the longest day is 13 and a half hours [13h 30']. Its second boundary is where the elevation of the pole over the horizon is 27 and a half degrees [27° 30'], and the longest day is 13 and three-quarter hours long [13h 45'].

There are around seven mountain ranges in this clime, a similar number of long rivers, and about 50 large cities known of. This clime begins in the east and stretches towards the west crossing central China, then it passes over the north of Sri Lanka, crosses northern Hind, the centre of the lands of Kabul, then through the lands north of Qandahār, the north of the lands of Sind, and the south of the lands of Makrān. Then it crosses the Persian Sea, passes through the middle of the lands of the Arabs, the Sea of al-Qulzum, the north of the lands of Abyssinia and the south of Upper Egypt, where it crosses the Nile, and then it traverses the middle of Ifrīqiya, the north of the lands of the Berbers, the south of the lands of Qayrawān, and the middle of Mauritania, to end in the Western Sea. The colouring of the majority of the people in these lands is between olive and black.

1 Manuscripts [ز and ك] give 600 miles.

List of the large cities in this clime, beginning in the East at the furthestmost extremity of China, that is every city whose latitude is from 20 [20°][1] to 27 degrees and 30 [minutes] [27° 30']. The first is in the East:

Names of the Cities	Longitude	Latitude
The city of *ṭ-w-ʿ-m-a* [?] in China	172° 40'	20° 45'
The city of *ṭ-ʿ-w-l-a* [?] in China	148° 10'	24° 30'
Al-Manṣūra in Sind[2]	98° 00'	22° 00'
Ṭā'if in Yemen	67° 10'	21° 00'
Mecca in Tihāma	67° 00'	21° 10'
Yathrib, the city of the Prophet	65° 00'	25° 00'
Akhmīm in Upper Egypt	55° 30'	27° 20'
Ifrīqiya in the West		

Table 3

[p. 173]

The Third Clime

The third clime is governed by Mars. Its longitude from east to west is 9,236 miles;[3] its latitude from south to north 350 miles.[4] Its boundaries are: 27 and a half degrees [27° 30'] and 33 degrees and 30 minutes [33°

1 *Sic.* As noted above, the first boundary of this clime is 20° 30'.
2 This is Brahmanabād, see Y. Friedmann, 'al-Manṣūra', in *EI2*, vol. 6, pp. 439–40.
3 Other manuscripts give: 6,000 miles [ج]; 200 [خ]; 6,255 [ر]; 7,200 [ز, ط, ف, ك, ل]; 8,200 [ن ,غ]; and 8,677 [marginal annotation in ن].
4 In other manuscripts: 300 miles [ع], and 3,500 [ف].

30']; its midpoint is where the elevation of the pole over the horizon is 30 and a half degrees and five [minutes] [30° 35']. The longest day is 14 full hours.

This clime has 33 mountain ranges, 22 long rivers, and 128 large cities known of. It begins in the east, passes through the north of the land of China and south of the lands of Gog and Magog, then it passes through the northern lands of Hind, the southern lands of the Turks, and then through the centre of the lands of Kabul, then the lands of Qandahār and Makrān, south of the lands of Sijistān,[1] and the middle of the lands of Kirmān; then it crosses [p. 174] Fārs (or Persia) on the seaboard, and passes the lands of Iraq to the south, then passes over the south of the lands of Diyār Bakr[2] and the north of the lands of the Arabs, then it traverses central Syria [al-Shām], the lands of Egypt [Miṣr], the lands of Alexandria, Marmarica,[3] the middle of the lands of Qādisiyya,[4] the middle of the lands of Qayrawān, then past the lands of Tangier, and ends in the Western Sea. The majority of the inhabitants of these lands are olive-skinned.

---

1 Sijistān is the Arabic form of Sistān. In the manuscripts containing this epistle both forms are used even within the same copy; see below p. 69.

2 Literally, 'the dwellings of the tribe of the Bakr', this is the designation of the region comprising the upper basin of the Tigris. The Turkish denomination, Diyarbakır, refers to the capital of the province, Āmid. Cf. M. Canard and Cl. Cahen, 'Diyār Bakr', in *EI2*, vol. 2, pp. 343–45.

3 Marmarica (*Marmārīqā* in Arabic) is the Latin name given to the North African region between Egypt and Cyrenaica.

4 There are several cities named Qādisiyya, here this toponym seems to refer to the Egyptian city bearing this name. Some manuscripts give *qāsiya* [ز, ل], *al-fāsha* [خ], and *al-fāsiya* [ط]. In al-Farghānī's description of this clime, Marmarica is not mentioned and the term listed between Alexandria and Ifrīqiya is Cyrenaica, lit. *bilād barqa*, see al-Farghānī, *Elementa Astronomica*, p. 37 of the Arabic text.

List of the cities in this clime, all with latitudes between 27 degrees and 30 minutes [27° 30'] and 33 degrees and 40 minutes [33° 40'].[1] The first is located in the East:

Names of the Cities	Longitude	Latitude
Qandahār in Hind	110° 00'	30° 00'
Kabul in Hind	100° 00'	33° 00'
Sirjān in Kirmān	107° 00'	32° 30'
Shīrāz in Fārs (or Persia)	73° 00'	30° 00'
Ahwāz in Khūzistān	75° 00'	32° 00'
Basra in Iraq[2]	74° 00'	31° 00'
Kufa in Iraq	69° 00'	32° 00'
Damascus in Syria [al-Shām]	60° 00'	38° 00'
Jerusalem in Palestine	56° 00'	38° 00'
Fusṭāṭ in Egypt [Miṣr]	35° 00'	29° 00'
Alexandria	59° 00'	31° 00'
Qayrawān in the West	31° 00'	32° 00'

Table 4

[p. 175]

1 This longitude, 33° 40', coincides with the one calculated by al-Farghānī, but has 10' less than the latitude given previously in this epistle for the second boundary of the third clime (33° 30'). Manuscript [ر] is consistent in this regard and gives 33° 30' in both cases.

2 MS [ٱ] mistakenly reads *al-Manṣūra* instead of *al-Baṣra*, which is the reading in all the other manuscripts.

The Fourth Clime

The fourth clime is governed by the Sun. Its longitude from east to west is 4,000 miles,[1] and its latitude from south to north 300 miles.[2] Its boundaries are 33 degrees and 30 minutes [33° 30'], and 39 degrees [39°], and its midpoint is where the elevation of the pole over the horizon is 36 degrees and 50 minutes [36° 50']. The longest day is 14 and a half hours long [14h 30'].

There are 25 mountain ranges, 22 long rivers, and approximately 212 large cities known of. This clime begins in the east and stretches through the north of the lands of China, the south of the lands of Gog, and to the south of the lands of the Turks, then to the lands of Hind; then it runs north of the lands of Balkh and Bāmiyān, then through the north of the lands of Makrān,[3] the middle of Sijistān,[4] central Fārs and Khūzistān,[5] the middle of the lands of Iraq, the middle of Diyār Rabīʿa[6] and then Diyār Bakr, the south of the lands of the frontier,[7] and the north of the lands of Syria [al-Shām]; then it crosses the centre of the Mediterranean Sea and the island of Cyprus, passes along the north coast of the lands of Egypt [Miṣr] and Alexandria, the north of Marmarica, the lands [fol. 38a] of Qādisiyya, Qayrawān, and Tangier, and it ends in the Western Sea. The colouring of most of the inhabitants in these countries is between olive-skinned and white.

This clime is the clime of the prophets and the wise men, because it is the central clime, with three climes to the south and three to the north and also because it is in the division of the Sun, the most brilliant body in the sky. The people of this clime have the most harmonious nature and equable morals. After them come the people of the two adjacent

1 Other manuscripts give: 300 miles [خ]; 7,800 miles [ك, ط, ز]; 800 miles [ع]; and 7,907 [margin of ن].

2 In manuscript [خ]: 466 miles.

3 The manuscripts give *kābilān* [أ, ج, غ], *īlān* [خ]. The province of Makrān is a semi-desert coastal region south of Sijistān.

4 i.e., Sistān; see above p. 67, note 1.

5 A province located in south-western Iran.

6 This refers to the eastern province of the northern lands between the Tigris and the Euphrates, whose capital was Mosul.

7 The Islamic-Byzantine frontier in Cilicia.

climes, I mean the third and the fifth. The inhabitants of the remaining climes, such as the Zanj, the Abyssinians, and most of the communities who inhabit the first and the second climes, lack the nature of the most excellent [clime], because their appearance is hideous and their behaviour savage. This is also the case for the communities in the sixth and seventh clime, such as Gog and Magog, the Bulgars, the Slavs, and similar peoples.

Names of the cities in this clime, with latitudes between 33 degrees and 40 minutes [33° 40'] and 39 degrees [39°] [p. 176]

Names of the Cities	Longitude	Latitude
Samarqand in Transoxiana (Warā' al-Nahr)	88° 20'	40° 00'
Balkh in Khurāsān	88° 35'	38° 40'
Herat in Khurāsān		
Marw in Khurāsān		38° 50'
Nīshāpūr in Khurāsān	80° 45'	37° 00'
Jurjān		
Āmul in Ṭabaristān		
Ray	75° 00'	35° 45'
Daylam in Jīlān		
Iṣfahān in Fārs	86° 00'	32° 25'
Hamadhān in the lands of Māhān	78° 00'	36° 00'
Baghdad in Iraq	70° 00'	38° 25'
Mosul in the Diyār Rabī'a	68° 10'	36° 30'
Aleppo in Syria [al-Shām]	68° 00'	35° 00'
Anṭākya	61° 34'	34° 10'
Amorium	58° 00'	40° 00'
al-Andalus in the West		

Table 5

The Fifth Clime

The fifth clime is governed by Venus. Its longitude from east to west is 4,000[1] miles and its latitude from south to north 255 miles. Its boundaries are from 39 [39°] to 43 and a half degrees [43° 30']; its midpoint is where the elevation of the pole is 41 and a third degrees [41° 20']. The longest day is 15 full hours long.

There are around 30 mountain ranges, 15 long rivers, and 200 [p. 177] large cities known of in this clime. It begins in the east and stretches through the middle of the land of Gog, the centre of the lands of the Turks, the lands of Farghāna and Isbījāb,[2] the middle of Soghdia and Transoxiana; then it crosses the Oxus, traverses the lands of Khurāsān, the north of Sijistān, Kirmān, then it traverses the northern lands of Fārs, and the centre of the lands of Ray and Māhān; then it passes through the northern lands of Iraq, the south of Azerbaijan, central Armenia, the north of the lands of the frontier, and the middle of Byzantium; then it crosses the Bosphorus,[3] goes along the northern Mediterranean, the middle of the lands of Rome, Port-Vendres,[4] and the central lands of al-Andalus, to end in the Western Sea. The majority of its inhabitants are white [in colouring].

Names of the cities of this clime with latitudes between 39 degrees [39°] and 43 degrees and 30 minutes [43° 30']

Names of the Cities	Longitude	Latitude
The lands of Gog and Magog		
The lands of the Khāqān of the Turks		
Ṭarāz in the lands of the Turks	87° 50'	39° 30'

1 This number is absent in manuscripts ج, خ, ر, غ, and ن. Other copies give: 400 miles [أ], and 7,400 miles [ز, ط, ك, ل].

2 Isbījāb, also written Isfījāb in Arabic, is the modern Sayram, in southern Kazakhstan.

3 Lit. Gulf of Constantinople.

4 Port-Vendres is a coastal town in the Oriental Pyrenees. It was called *Portus Veneris* by the Romans, after the goddess Venus, who had a temple there, whence the Arabic toponym *haykal al-zuhara*, 'Temple of Venus'.

Names of the Cities	Longitude	Latitude
Khwārizm in Transoxiana		
Ardabīl in Azerbaijan		
Khilāṭ in Armenia	65° 00'	39° 40'
Malaṭyā in Armenia	61° 00'	39° 00'
Macedonia[4]	49° 00'	40° 00'
Greater Rome of the Romans[5]	35° 10'	41° 50'
Port-Vendres	59° 30'	42° 10'

Table 6

The Sixth Clime

The sixth clime is governed by Mercury. Its longitude from east to west is 5,000 miles,[1] and its latitude from south to north 216 miles.[2] Its boundaries are from 43 and a half degrees [43° 30'] to 47 and a quarter degrees [47° 15']; its midpoint is where the elevation of the pole is [p. 178] 45 degrees and fifty minutes [45° 50']. The longest day is 15 and a half hours long [15h 30'].

There are around 22 mountain ranges in this clime, 32 long rivers, and 90 large cities known of. It begins in the east and traverses the north of the lands of Gog, the south of Sijistān, the lands of the Toghuzghuz,[3] the middle of the lands of the Khāqān,[4] the lands of the Kīmāk,[5] the north of Soghdia and Transoxiana, the middle of Khwārizm, the north of Isbījāb, the north of Jurjān, Ṭabaristān, Daylam and Jīlān;[6] then it crosses the Sea

1 Other manuscripts give: 7,000 miles [ز, ط, ل], and 700 miles [ك].
2 Other manuscripts give: 200 [ج, خ, ل, ن]; 210 [ز, ك]; and 220 miles [ف].
3 The Toghuzghuz were a nomadic Turkish tribal confederation who lived in the northwest region of modern China; see P. B. Golden, 'Toghuzghuz', in *EI2*, vol. 10, pp. 555–57.
4 Khāqān was the honorific title of the ruler of the Khazar Turks.
5 The Kimäk, or Kīmāk in Arabic, were a Turkish people living in what is now Kazakhstan; see Bosworth, 'Kimäk', in *EI2*, vol. 5, pp. 107–108.
6 Jīlān was a small province lying west of the southern Caspian; see G. Le Strange, *The Lands of the Eastern Caliphate* (Cambridge: Cambridge University Press, 1930), p. 172ff.

of Ṭabaristān,[1] runs through central Azerbaijan, through the middle of Armenia and Malaṭyā, north of the Pontus,[2] north of Constantinople, the middle of Macedonia, the middle of the lands of Ifrīqiya[3] towards the north, passing by the south of the Sea of the Slavs and the north of Port-Vendres, and ends in the Western Sea. The colouring of most of the inhabitants in these lands is between fair and white.

Names of the cities in this clime with latitudes between 43 degrees and 30 minutes [43° 30'] and 47 degrees and 15 minutes [47° 15']. It starts in the East (fol. 38b)

Names of the Cities	Longitude	Latitude
The lands of Gog	87° 00'	42° 00'
The lands of the Khazars	98° 00'	45° 00'
Island in [the Sea of] Jurjān		
Bardhāʿa in Azerbaijan	78° 00'	43° 00'
Darband [Bāb al-Abwāb]		
Constantinople of the Romans [Qusṭanṭīniyya]	49° 50'	45° 00'
The lands of the *B-r-j-ā-n* [?][4]	40° 00'	45° 00'
The Land of Afīlyūn[5]	25° 45'	44° 55'

Table 7

1 This is the Caspian Sea.

2 i.e., the Black Sea.

3 Ifrīqiya is the Arabic toponym that designates the area of Northern Africa which corresponds to modern Libya, Tunisia, and Eastern Algeria. The capital of Ifrīqiya was Qayrawān, which is now in central Tunisia.

4 The manuscript reads *bilād b-r-j-ā-n*: the Burgundians?

5 The table in the manuscript reads *bilād afīlyūn* (while in the body of the text it is *al-w-y*). The term refers to Albion, the Greek name for Great Britain. The *Alwy* of the text is presumably to be vocalized as *Alūya*, cf. D. M. Dunlop, 'The British Isles According to Medieval Arabic Authors', *The Islamic Quarterly*, vol. 4 (1957), pp. 11–28.

The Seventh Clime

The seventh clime is governed by the Moon. Its longitude from east to west is 6,600 miles and its latitude from south to north is 185 miles. Its boundaries range from 47 and a quarter degrees [47° 15'] to 50 and a half degrees [50° 30'] and its midpoint is where the elevation of the pole over the horizon is 48 and two thirds [48° 40'].[1] The longest day is 16 full hours.

There are around ten mountain ranges in this clime, approximately 40 long rivers, and about 22 large cities known of. It begins in the east and stretches towards the south of the lands of Gog and Magog, Sīstān, then the lands of the Toghuzghuz, the lands of the Kīmāk, the south of the lands of the Alāns,[2] the north of the Sea of Jurjān, the land of the Khalaj,[3] then Mount Darband;[4] then it crosses the Pontus, passes through [p. 179] the south of the Sea of the Slavs and the south of the Island of Alūya,[5] and ends in the Western Sea. Most of the inhabitants of these lands are fair in colouring.

These are the cities which have a latitude from 47 degrees and 15 minutes [47° 15'] to 49 degrees [49°]. It begins in the east.

Names of the Cities	Longitude	Latitude
The lands of Gog and Magog		
The lands of the Alān Turks		
An island in the Western Sea		

1 Other manuscripts read 48° 50' [ر] and 48° 30' [ع, ن].

2 The Alāns were an Iranian people who inhabited the north Caucasus; see P. B. Golden, 'Alāns', in *EI3*.

3 The Khalaj were a pastoralist Turkish tribe from the Central Asian steppes; see G. Doerfer, 'Khaladj', in *EI2*, vol. 4, pp. 917–18.

4 In Arabic *bāb al-abwāb*, literally, 'the gate of gates' or 'the great gate'. Darband was also known as the Pass of the Caucasus, the Caucasian Gates, and the Gate of the Alāns or Darial Pass (*Bāb al-Lān* in Arabic; *Dār-i Alān* in Persian); see C. d'Encausse, 'Derbend', in *EI2*, vol. 2, p. 206.

5 Other manuscripts offer different readings of this toponym: *al-Rayy* [ر, خ], the reading of the mediaeval Latin translation (*insula Ray*); *al-barbar* [غ], and *al-l-w-n* [ف]. On this term see previous page, note 5.

Names of the Cities	Longitude	Latitude
An island		
An island		

Table 8

Chapter

You should know, O Brother, that there are thousands of cities in each of these seven climes. Their number changes constantly. In each city there are communities of people dissimilar to each other in terms of language, colouring, nature, behaviour, opinion, way of life, occupations, crafts, and customs. This is also true of animals and mineral resources, which vary in form, taste, colour, and smell. The reason for this is the differences between the air of each country, the soil in the fields, and the freshness and salinity of the waters. These differences are entirely dependent on which zodiacal signs are in the ascendant and on their degrees above the horizon [p. 180] of any given country, on the course of the planets over the pastures under them, and on where they cast their rays.[1] It would take a long time to explain this. It is said that one of the ancient kings once ordered that the cities of the inhabited quarter of the Earth be tallied. He discovered that there were a little over 17,000 cities, to say nothing of the villages.

You should know that the number of the cities of the Earth increases and decreases. This depends on what is determined by the rulings of the planetary conjunctions and the revolutions of the spheres. The planetary conjunctions indicate the power of the fortunate stars, the harmoniousness of the era, the equilibrium of the basic elements, the advent of the prophets and the continuation of revelation, the abundance of scholars, the justness of kings, the well-being of the human estate, and the descent of heavenly blessings in the form of rain, so that the earth and the plants grow and bear fruit, the animals

1 The influence of the rays of the stars had been treated by al-Kindī (d. 256/873) in his *De Radiis*; see P. Travaglia, *Magic, Causality and Intentionality: the Doctrine of Rays in al-Kindi*, Micrologus Library 3 (Florence: Galluzzo, 1999). Also P. Adamson, *Al-Kindī* (Oxford: OUP, 2007), p. 188f.

procreate and multiply, countries prosper, and cities and civilization increase. The planetary conjunctions [also] indicate the power of the unfortunate stars, the corruptness of the age, the disequilibrium of the mixture of humours, the cessation of revelation, the paucity of scholars, the termination of good deeds and royal munificence, the corruption of morals, the wickedness of actions, disagreements, and the withholding of heavenly blessings in the form of rain, whereby the land does not become fruitful, plants wither, animals perish, and cities fall into ruin.

Chapter

You should know, O Brother, that the matters of this world are subject to cyclical upheavals[1] and misfortunes century after century, from community to community, and from country to country. Know that every state has a beginning in time, a zenith towards which it ascends, and a point at which it ends. When it attains its apogee and reaches its furthest extent, it starts to descend and decay. Its inhabitants cease to co-operate and begin not to prosper. Strength, energy, hegemony, and prosperity appear amongst other people. The latter gain in strength and grow while the former begin to weaken and decrease until the first people, the ones who prevailed earlier, come to naught, and the second group, the ones who arrived later, establish themselves.

The regular passage of time illustrates this. It is divided into bright day [p. 181] and dark night. It is further divided into hot summer and cold winter. These two seasons revolve and follow each other as they come and go. When one arrives, the other leaves. Now one increases and now the other decreases.

Whenever one decreases, it effects a proportionate increase in the other, so that when they both reach their limits in terms of increase and decrease, the one that has ceased to increase begins to decrease, and the one that has ceased to decrease begins to increase. They continue

1 *Dawla* is difficult to translate. The Arabic root *d-w-l* denotes change and the substantive *dawla* may mean revolution, upheaval, turn and, by extension, also reign, rule, or dynasty.

thus until they are equal in their extent, and then they exceed their respective states and advance to the limit of their increase and decrease. Whenever one of them reaches its limit of increase, its strength becomes manifest and its effects predominate in the world, whilst the strength of its opponent becomes imperceptible and its effects few.

Such too is the order of time when it comes to the cycle of the virtuous and the cycle of the evil-doers. At one time the cycle, the power, and the prevalence of effects in the world are in favour of the virtuous; at other times the cycle, the power, and the prevalence of effects in the world are in favour of the evil-doers (fol. 39a). As God, May He be Praised and Glorified, says: *such days We give to men by turns;*[1] *but none understands them save those who know.*[2]

Chapter

Dear, esteemed Brother, may God aid you and us with a spirit of His, we observe that in this epoch the cycle of the evil-doers has reached its peak, their power has become manifest and their actions predominate in the world. But after reaching the limit there is only descent and decay. You should understand that power and kingship alternate in every epoch, time, cycle, and conjunction, from community to community, household to household, and country to country. And know, O Brother, that the inception of the cycle of the virtuous appears with a group of good, wise, virtuous, and learned scholars. They are unanimous in thought and acknowledge one religion only. They have taken a pact and sworn an oath of fidelity not to dispute one with another,[3] and not to refrain from helping one another, to be as one man in all matters, and as a single soul in [p. 182] all their dealings, in furtherance of their aim to support their religion and to seek the afterlife. They seek naught

1 Qur'an 3:140.

2 Qur'an 29:43.

3 Theological disputation [*jidāl*] was a common and important practice in early ʿAbbāsid intellectual life. However, the debates conducted by the 3rd/9th and 4th/10th-century theologians were regarded by many scholars as a threat to Islam. This denunciation of *jidāl* is thus both an intellectual desideratum and an ideological stance.

but the face of God, sublime is His praise and favour, [who *desire not*] *recompense, not thankfulness.*[1]

Dear, esteemed Brother, may God help you and us with a spirit of His. Is there then any good reason not to prefer the company of brothers who are of wise counsel and are friends: good men and true — for such is their description? Is there any reason for you not to seek what they seek, adopt their morals, study their sciences, learn their methods, and be with them and so be saved *in their security; evil shall not visit them, neither shall they sorrow?*[2]

May God guide you, O Brother, and all our brothers, to the right path with His grace and His favour; He is the Governor of this [world]. All lies within His power.

****

The fourth epistle, the epistle on geography, has come to an end. Praise be to God. It is right to praise Him. Blessings on our Prophet Muhammad, his family, and companions.

1 Cf. Qur'an 76:9; the entire verse reads: *We feed you only for the Face of God; we desire no recompense from you, no thankfulness.*

2 Qur'an 39:61; the entire verse reads: *But God shall deliver those that were god-fearing in their security; evil shall not visit them, neither shall they sorrow.*

Appendix 1

Comparison of the Description of the Climes in the text of the *Epistle on Geography* of the Ikhwān and the *Compendium* of al-Farghānī[1]

	Ikhwān	Al-Farghānī
First Clime	Island of Rubies, China, Sri Lanka, Hind, Sind, the Persian Sea, Oman, Abyssinia, the Egyptian Nile, Nubia, the land of the Berbers, Libya, Mauritania, the Western Sea	China, the city of the king of China, *Ishfitīra*, Hind, Sind, the Island of Kark, the Arabian Peninsula, Yemen (which contains the cities of Ṭafār, Oman, Hadhramawt, Aden, Sanʿa, al-Qayn, Mārā, Tabāla, Jarsh, Mahr, Sabā), the Red Sea, Abyssinia, the river Nile of Egypt, an Abyssinian city named Jarma, Dongola in Nubia, the land of the Berbers, the Western Sea
Second Clime	China, Sri Lanka, Hind, Kabul, Qandahār, Sind, Makrān, the Persian Sea, the lands of the Arabs, the Red Sea, Abyssinia, Upper Egypt, the Nile, Ifrīqiya, the lands of the Berbers, Qayrawān, Mauritania, the Western Sea	China, Sind, Brahmanābād, al-Bīrūn, Daybul, the Indian Ocean, the Sea of Basra, the Arabian Peninsula, the Najd, Tihāma, (which contains the cities of al-Yamāma, Bahrayn, Hajr, Yathrib, the Ḥijāz, Mecca, Ṭā'if, and Jidda), the Red Sea, Upper Egypt, the Nile (passing by the cities of Qūs, Akhmīm, *a-n-s-n-ā* (?), Antinopolis, Aswān), Ifrīqiya, the lands of the Berbers, the Western Sea

1 From al-Farghānī, *Elementa Astronomica*, ed. J. Golius (Amsterdam: Jasonius à Waasberge, 1669).

Third Clime	China, the lands of Gog and Magog, Hind, the lands of the Turks, Kabul, Qandahār, Makrān, Sijistān, Kirmān, Persia (Fārs), Iraq, Diyār Bakr, the lands of the Arabs, Syria (al-Shām), Egypt, the lands of Alexandria, Marmarica, Qādisiyya, Qayrawān, Tangier, the Western Sea	China, Hind, Qandahār, Sind, Kabul, Kirmān, Alexandria (*sic*), Sijistān, al-Maḥmūdiyya, Jiruft, Sirjān, the coast of the Sea of Basra (with the cities of Iṣṭakhr, Jawr (?), Fāsā, Sābūr, Shīrāz, Siraf, Janābā, and Mahrūbān), Bakūr al-Ahwāz, Iraq, Basra, Wāṣit, Kufa, al-Anbār, Hīt, Syria (al-Shām) (containing the cities of Ḥiyār, Salamiyya, Homs, Damascus, Ṣūr, Acre, Ṭabariyya, Caesarea, Jerusalem, Arsūf, Ramla, Asqalan, Gaza, Midian, Clysma), Egypt (containing Farma, Tanis, Damietta, Fusṭāṭ of Miṣr, Faiyum, and Alexandria), Cirenaica, Ifrīqiya, Qayrawān, the Western Sea
Fourth Clime	China, the lands of Gog, the lands of the Turks, Hind, Balkh, Bāmiyān, Makrān, Sijistān, Fārs, Khūzistān, Iraq, Diyār Rabīʿa, Diyār Bakr, the lands of the frontier (*bilād al-thaghr*), Syria, the Mediterranean, Cyprus, Egypt, Alexandria, Marmarica, Qādisiyya, Qayrawān, Tangier, the Western Sea	Tibet, Khurāsān (containing the cities of Khujand, Usrūshana, Farghāna, Samarqand, Balkh, Bukhara, Herat, Umwayh, Maywand, Marr, Sarakhs, Ṭūs, Nishāpūr, Jurjān, Qūmis, Ṭabaristān, Dīnawānd, Qazwīn, Daylam, Ray, Iṣfahān, Hamadhān, Nihāwand, Dīnawar, Ḥulwān, Shahrazūr, Samarra, Mosul, al-Balad, Nusaybīn, Āmad, Ra's al-ʿAyn, Qālīqalā, Arsamosata, Ḥarrān, Raqqa, Qarqīsiyya), Syria, Balis, Manbij, Samosata, Malaṭyā, (al-z-n-ṭ-r-a ?), Aleppo, Qinnasrīn, Antioch, Tripoli, Maṣīṣa, Sidon, al-Kanīsa al-Sawdā', Adana, Tarsus, Amorium, Latakya, the Sea of Syria, Cyprus, Rhodes, Tangier, the Western Sea

Fifth Clime	The lands of Gog, the lands of the Turks, Farghāna, Isbījāb, Soghdia, Transoxiana, the Oxus river, Khurāsān, Sijistān, Kirmān, Fārs, the lands of Ray, Māhān, Iraq, Azerbaijan, Armenia, the lands of the frontier, Byzantium, the Bosphorus, the Mediterranean, Rome, Port-Vendres, al-Andalus, the Western Sea	Gog, Khurāsān (containing the cities of Ṭirāz, Nawakāth, Khwārizm, Isbījān, Shāsh, Trebizond), Azerbaijan, Armenia, Bardhāʿa, Nashawā, Sīstān, Arzān, Khilāṭ, Byzantium, Kharshana, Qurra, the Great Rome, the Sea of Syria, al-Andalus, the Western Sea
Sixth Clime	The lands of Gog, Sijistān, the lands of the Toghuzghuz, the lands of the Khāqān, the lands of the Kīmāk, Soghdia, Transoxiana, Khwārizm, Isbījāb, Jurjān, Ṭabaristān, Daylam, Jīlān, the Sea of Ṭabaristān, Azerbaijan, Armenia, Malaṭyā, the Pontus, Constantinople, Macedonia, Ifrīqiya, the Sea of the Slavs, Port-Vendres, the Western Sea	Gog, lands of the Khazar, the Sea of Jurjān, Byzantium, Kharzān, Amasya, Heraclia, Chalcedon, Constantinople, the Land of the Burjān,[1] the Western Sea
Seventh Clime	The lands of Gog and Magog, Sīstān, the lands of the Toghuzghuz, the lands of the Kīmāk, the lands of the Alāns, the Sea of Jurjān, the lands of the Khalaj, Mount Darband, the Pontus, the Sea of the Slavs, the Island of Alūya, the Western Sea	Gog, the lands of the Turks, the Sea of Jurjān, the Mediterranean, the lands of the Burjān, the lands of the Slavs, the Western Sea

1 This term may refer to the Burgundians.

Appendix 2

Co-ordinates of the cities of the seven climes compared with al-Khwārizmī's values

The following tables reproduce the co-ordinates of the cities listed in the *Epistle* as they occur in the manuscripts, in the majority of instances without diacritics. The left column contains the co-ordinates given in the *unicum* manuscript of al-Khwārizmī's *Kitāb Ṣūrat al-arḍ* edited by Mžik.

Major cities of the seven climes	Manuscripts of the *Epistle on Geography*	Al-Khwārizmī
Kashmir in Hind	,[ط] قز، يا د ,[ر] بو د، باد ,[خ] بو د، بار فلو، ,[ل] قكز يا ,[ف] فلر، فلر ,[غ]فكر، ح [ن]فلر	
Island of Kark in Hind	قب كہ ,[ط] قب، طہ ,[ر] فب، ط ,[خ] فب، ط قب، ط ,[ف] قب، ط ,[غ] ص عب، ط ,[ع] ط [ن] عب، ط ,[ل]	
The City of Perfumes in Hind	عب، ه,[ر] عب، ه ,[خ] عب، ه ,[أ] — ، د ل ,[ف] عب، ه ,[غ] عب، ه ,[ع] عب، ه ,[ط] [ن] عب، ه ,[ل]عبه، -	عب ♉، د ل
Hadhramawt in the Yemen	عا، ,[ر] عا، ن ,[خ] عا، ن ,[أ] عا ♉، ند، ♉ ,[ف] لح، نا ,[غ] عا، ب ,[ع] نب عا، ,[ط]يد [ن] عا، ب ,[ل]نح، نا	عا ♉، ب
Waghāra in Abyssinia	س، ,[ط] س، ل ,[ر] س، نب ,[خ] س، ن [ن] عا، ب ,[غ] لح، نا ,[ع]نا	س نه، نا ♉
Kūkū in Abyssinia	نح، ,[ر] نح، نا ,[خ] نح، نا ,[أ] ـ، ه ♉ [ن] لح، نا ,[ل] نح، نا ,[ع] لح، نا ,[ط]ما	مح ♉، ى نه

Cities of the first clime	Manuscripts	Al-Khwārizmī
A city in furthermost China	فعد ل، عح به [ر], فعد، مح ل [خ], فعد، طل فعد ل، [غ] نعو ل، نح مه [ع], قعه، لح [ط], [ك] سد، ط ط [ف], لح به	
A city in the islands of China	قم، [ط] قعب، يو [ر], فعب، لو [خ], فعب، لو [ك] فل، ب [غ], سف، لط [ع], نر	
The city of *ʾ-s-?-r-?-ā-r* (?) in China	سد، ط [ر], فعد، ط ط [خ], فسد، ط ط سد، ط [غ], فكر نا، لر به [ع], بدح، يط [ط], ط [ك] فكه، نو به [ف], ط	فسر به، يه ل
B-ā-y-w-q-a (?) in Hind	فكه، [خ], فكه، يب [أ], فل ♉، ب ♉ قكه، [ع], قد، م [ط], قل، يب [ر], يب [ك] في، بو [ف], فل، فب [غ], لو	
Khaberis in Hind	[ط], فكه، يو به [ر], فى، نو به [خ], فى، نو به [ك] فد ل، مط به [ف], فكه، بو به [ع], قكه، يو	
M-s-t-ʿ-l-y (?) in Hind	مر، [ط] قي، يو [ر], فدل، لد [خ], فول، لو [ك] مه، بج [ف], فى، بو [غ], ح، لط مه [ع], يو	ق ♉، يه ل
Oman in the land of the Arabs	فد ك، مط [ر], فه، مط به [خ], فه، به بط فدك، مط [غ], مد ل، ل ح [ع], قد، بط [ط], به [ك] سد سد، بح [ف], به	فد ل، نط مه
The city of *al-m-y-d* (?) in Hind	فد، لح [ط], فه، لح [ر], ق، بح [خ], ق، بح [ك] سد، بد [ف], له، لح [غ], فه، له [ع],	فد ل، ب ♉[1] فر ♉، ط ♉ فط ♉، يج ل
Another city	[ع], فد، يه [ط], ب، به [ر], لح، به [خ], بح، به [ك] بح، به [ف], ب، به [غ], و، -	
Ẓafār in the land of ʿĀd	سه، [غ], سفر، به [ع], سد، ه [ط], سه، بج [ف], لحـ	عح ♉، به ♉
Aden in the land of Yemen	[ر], سد، فح [خ], سد، بح [أ], سه ل، بحـ ♉ سد، [غ], لح، به [ع], سد، ح [ط], سد، يح [ك] لك، ب [ف], لح	سه ♉، بح ♉
Sanʿa in Yemen	[ف] سد، بد [غ], به، مب [ط], سد، يد	سح ل، بد ل

1 Al-Khwārizmī lists other two unnamed cities in the island of al-Maydh with the following co-ordinates: فر ♉، ط ♉, and فط ♉، يج ل.

Dongola in Nubia	بح، [ر], بح، به [خ], بح، به [أ], م نج، بد ♉ [ف], لج، بو [غ], مب، ب [ع], لح، يو [ط], به [ك] بت، ك	نج بد، -
Kusti	[ك] لب، ك [ف], ل، ب [ع], د، م [ط], ل، نب	
The kingdom of Abyssinia	مد، [ط], نب، ك [ر], ب، بحو [خ], ب، بحو [ل], [ك], ك بط [ف], ب، ك [غ], فد، ك [ع], يط	
The Greater Jarma	[ع], لو، كه [ط], لب، ك [ر], لد، ل [خ], لو، لط [ك], كو، بط [ف], لب، ك [غ], كو، ك	لد ♉، بط ل
The lands of the Berbers	كو، [ط], ك يط [ر], بحو، بط [خ], نحو، بط [ف] ك بط [ع], بط	
Libya	[ف] كو، بط [ط], كو، بط	

Cities of the second clime	Manuscripts	Al-Khwārizmī
The city of *ṭ-w-ʿ-m-ā* (?) in China	قمد، ك [ر], مو، كر [خ], فد، ك [أ], قمب م، كمه [ك] فمد، ك [ع], قد، كر [ط], مه	
The city of *ṭ-ʿ-w-l-ā* (?) in China	[ر], قح، كو ل [خ], فح، كو ك [أ], قمح ى، كد ل [ك] بح، مه [ع], لح، كد [ط], قمب، كد ل	فمح ى، كد ل
The city of *ʾ-ṭ-w-ā-s-ā* (?) in China	[ك] فد، كد [ع], قم، ل [ط], قم، اكد [خ], فح، دلد	قم ل، كد ♉
Quni in Hind	فد، كد [ط], قد، كد [ر], فد، كا [خ], قد، كا [ك] فا بح، له [ع]	قد ى، كد مه
A city	به بح، [ع], [ط], قيا، كز [ر], فما، كو [خ], قما، كو [ك] اكد	
Al-Manṣūra (Brahmanabād) in Hind	بح، [خ], فح، لب [أ], فح ♉، كب ♉ فحـ ♉، [ع], صح، كد [ط], محب، كب [ر], اكب [ك] نب لب، اك [غ], كب ♉	صدك، كح ل
Al-Bīrūn by the sea	مب، كجـ [ر], فسب، سح ل [خ], فسب، لح ل سح، لح [غ], قب ك، كو ل [ع], صد، كح [ط], ل [ك] ل	صب ك، كح ل
Daybul	[ك] مه عا، كر لب [ع], صب، كج [ط], كب، كدك	صب ♉، كدك
Diyār Bakr	[ك] سه، كد م [ع], مح، كد [ط], محـ مه، كد م	

Yamāma in the Ḥijāz	عا مه، كا, [ر] عا م، كا ل, [خ] عا مه، عا ل [ك] سر، له ك, [غ] فا به، كا ل, [ع] عا، كا, [ط] د	عا مه، كا ل
Ṭā'if in Yemen	[ر] سد، كو, [خ] سه، كو, [أ] سر ب، كا ♉ [ك] ل، لب, [غ], [ط] سح، كب	سح ك، كا ك
Mecca in the Tihāma	به، [ع] س، كا, [ط] سز، كا, [أ] سر ♉، كا ك [ك] كا م	سر ♉، كا ♉
Yathrib, the City of the Prophet	[ر] سه، كو, [خ] سه، كو, [أ] سه ♉، كه ♉ ل، [غ] عر، كا م, [ع] سه، كد, [ط] سه، كه [ك] له, [ل]	سه ك، كه ♉
Akhmīm in Upper Egypt	له، [ع] يه، كو, [ط] نه ل، كول, [أ] نه ل، كر ك [ك] كا	نه ل، كو ن
Ifrīqiya	ب، لو ك, [ط] له، كو, [ر] لح، كو, [خ] لح، كو [ك], [ل]	
Libya	ك، لو, [غ], [ط] كه، كا, [ر] مح، كو, [خ] مح، كو [ك] ل	
Ma'd in the Tihāma	[ر] سر، كا, [خ] سر، كا	

Cities of the third clime	**Manuscripts**	**Al-Khwārizmī**
The city of *ʾ-r-a-b-d-a* (?) in China	[ر] فد، لح ل, [خ] قسد، كح, [ج] فس، لح ل [ك] فخ فك، لك لج, [ف] فر، عح ل, [ع] قس، لح	فس ♉، كح ل
Qandahār in Hind	[خ] فى، ل, [ج] فى، ل, [أ] قى ♉، ل ♉ فى، [ع] فى، ل, [ط] قيجـ، ل ك, [ر] فى، ل [ك] صد به، كلـ, [ف] ل	قى ♉، ل ♉
Kabul in Hind	[خ] ن، كح, [ج] ن، كحـ, [أ] ق ♉، عح ♉ ن، كح, [ع] ق، كح, [ط] قك، كح, [ر] ف، لح [ك] صب، لط ب, [ف]	ق ♉، كح ♉
Zaranj in Sijistān	[ر] صد به، -, [خ] صد به، -, [ج] ط به، لب ك [ف] مد، به, [ع] سد، كد, [ط] صد نه، لب ك [ك] يح ع، لد ل	
Ṭīz in Makrān	صب م، [خ] صب م، بكطل, [ج] صد م، كط ل صب، [ط] صب م، كط ل, [ر] من كطل [ك] عه م، لط لو, [ف] مح، لد د, [ع] كظ	ف م، كط ♉

Sirjān in Kirmān	يح، ا ب د ,[ج] مح، لد د ,[أ] قح ♉، لب ل قح، ,[ط] فج، لب ز ,[ر] يح، لب د ,[خ] [ك] -، لد لا ,[ف] مح، لد د ,[ع]لب	فح ♉، لب يه
Shīrāz in Fārs	[خ] عج، ل ,[ج] كح، ل ,[أ] عح ♉، ل ♉ [ع] عخ، لب ,[ط] عح، كط لو ,[ر]عح، كط لو [ك] س، لب ,[ف]عح، ل	عح ♉، لب ♉
Ahwāz in Khūzistān	عه، ,[ج] عد، لب ,[أ] عه ♉، لب ♉ [ع] عو، لب ,[ط] عه، لب ,[ر] عه، ل ,[خ]ل [ك] يو، لح ,[ف]عه، لد	عه ♉، لب ♉
Basra in Iraq	سط ل، لا ن ,[ج] عد، لا ,[أ] عد ♉، لا ♉ عد، ,[ع] عد، لا ,[ط] عد، لا ,[ر] عد، لا ,[خ] [ك] لر م، لب ,[ف]لا	عد ♉، لا ♉
Kufa in Iraq	سط ,[ج] سطر، لا ر ,[أ] سط ♉، لب ♉ يط ,[ع] سط، لا ,[ط] سط ل، لا نه ,[ر]ل، لا ن [ك] س، كط ,[ف] ل، لا ب	سط ل، لا ن
Damascus in Syria	[خ] س، لح ,[ج] س، لح ,[أ] س ♉، لح ♉ س، ,[ع] س، لج ,[ط] س، لج ,[ر]س، لح [ك] يه، يه لا ,[ف]لح	س ♉، لح ♉
Jerusalem in Palestine	[خ] يب، لب ,[ج] يد، لب ,[أ] يو ♉، لح ♉ ير، لد,[ع] يو، لب ,[ط] نو، لب ,[ر]يد، لب [ك] لا يه، مه لا ,[ف]	
Fusṭāṭ in Egypt (Miṣr)	يو م، لحط ,[ج] يد م، كله ر ,[أ] نه ♉، كط ♉ [ط] لز م، كط يه ,[ر] يب م، كط به ,[خ] به [ك] ك، له ,[ف] يد م، كط يه ,[ع]يب، لا لا	
Alexandria of Ptolemy	[خ] سب، لا ,[ج] سب، لا ,[أ] نط ♉، لا ♉ [ف] سـ، لا ,[ط] س، لا ,[ر]سب، لا	نا ك، لا ه
Qayrawān in the West	لا نه، لا ,[ج] لا يه، لا مه ,[أ] لا ♉، لب ♉ لا يه، ,[ط] لا مه، لا مه ,[ر] لا يه، لا يه ,[خ]نه [ك] -، لح ,[ف]لا يه	لا ♉، لا م
Tangier in the West	ما ح، عط ,[خ] يا ح، عط لب ,[ج] يا ح، عط لد [ف] ما ح، عط لر ,[ع] يا، كط ,[ط]ك، له ,[ر]لر	ح ♉، له ل
K-?-a-n (?)	[ط] صه، لج	

Cities of the fourth clime	**Manuscripts**	**Al-Khwārizmī**
The city of *k-a-sh-ʿ-r* in China[1]	[ك] قف ص، كط	
Tibet in the lands of the Turks	قف، كط, [ر] مد، لح, [خ] لو، يح, [ج] ند، يح [ك] صد ك، لط, [ف] ند، يح, [ع] فد، لح, [ط]لح	فل ♉، لح ♉
Samarqand in Transoxiana	[خ], لط، لون, [ج] نك، لون, [أ] فخ ك، م ♉ نط، لون, [ع] قد، لو, [ط] ص، لط, [ر]لط، لون [ك] فد مه، لو ر, [ف]	فط ل، لر ل
Balkh in Khurāsān	لح ل، لح, [ج] لح ل، لح م, [أ] فخ له، لح م نح ل، لح, [ع] قد، كد, [ط] صد ك، لو ن, [خ]م [ك] ومه، نو, [ف]م	فج له، لح م
Herat (al-Harrāh) in Khurāsān	قد، , [ط] فه، لو, [ر] قدك، لون, [خ] فدك، لون [ك] و، لج, [ف]ا مه، لو, [ع]كد	
Marw in Khurāsān	لب, [خ] لد نه، لو, [ج] ند ل، لح ن, [أ] _، لح ن [ك] كعد، لو, [ط] فد، لح, [ر] نه، لو	فد ك، لح له
Nishāpūr in Khurāsān	ف، مح له, [خ] ف، نح ل, [أ] ف مه، لر ♉ عه، لح, [ف] و، لح ل, [ط] ف مه، لو, [ر] [ك]لو	ف مه، لز ♉
Jurjān	[ط], ف، لح ن, [خ] لوع، لر نه, [ج] ر نه، لر [ك] عد م، لر مه, [ف] عد د، لر نه, [ع]	ف مه، لح ن
Āmul in Ṭabaristān	عز ك، لر مه, [ر] عه، ل, [خ] عه، ل, [ج] ں، لح ل عح لو، لر, [ف] عه، ل, [ع] عو، لو, [ط] [ك]مه	عز ك، لز مه
Ray in Fārs	[خ], عد، لح ن, [ج] عو ل، لر مه, [أ] عه، له مه عد، , [ع] عه، له, [ط] عد، لد مد, [ر]عد، لح ب [ك] سط، لح لب, [ف]لح ر	عه ♉، له مه
Daylam and Jīlān	[ر], عد م، لد ن, [خ] عد م، لو ن, [ج] عه، له نا، , [ف] عد م، لو ں, [ع] عه، لح, [ط]عه، لح ى [ك]كه لب	عه ♉، لح ى[2]
Iṣfahān in Fārs	[خ], عح، لو, [ج] عد، لح ر, [أ] فو ♉، لب كه عح، , [ع] عد، لد, [ط] عد م، لب كه, [ر] عج، لو [ك] له، عج, [ف]لو	عد م، لد ل

1 This toponym may correspond to Ptolemy's Kattighara.

2 The co-ordinates correspond to Daylam.

Hamadhān in the Land of Māhān	ع، ,[ج] عد م، لد ن ,[أ] عحـ ♉، لو ♉ ع،,[ع] كح، لو ,[ط] عجـ، لو ,[ر] ع، لح ,[خ]لح [ك] سو، له ک ,[ف]لح	عح ♉، لو ♉
Baghdad in Iraq	سط، له ,[ج] عح، لو[1] ,[أ] ع ♉، لحـ كه ,[ع] ع، لح ,[ط] ع، لجـ ,[ر] سط، لب ل ,[خ]ل [ك] يح، لج ,[ف] سط، لدل	عح ♉، لح ط
Mosul in the Diyār Rabīʿa	,[خ] سح، له ل ,[ج] عح، لو ,[أ] سح ى، لو ل سح، ,[ع] سط، له ,[ط] سط، له ک ,[ر]سح، له ل [ك] ى، لول ,[ف]له ل	سط ♉، له ل
Aleppo in Syria	سا له، لو ,[ج] ع، لح ,[أ] سحـ ♉، له ♉ سح، ,[ط] سا له، له ي ,[ر] سا ل، لو ى ,[خ]ى [ك] ک، فى د ,[ف] سا له، له فى ,[ع]له	سح ♉، لد ل
Cyprus	س، ,[ط] سو، لول ,[ر] ع، لح ,[خ] ع، لح [ك] -، لح ,[ف] ع، لح ,[ع]لو	
Anṭākya	,[ر] س ر، لو ,[خ] سر، لو ,[أ] سا لد، لد ى [ك] عد، فى ,[ف] مر، لو ,[ع]سل، له	سا له، لد ل
Amorium	لح، ,[ر] لو، ع ,[خ] لو، ع ,[أ] نخ ♉، م ♉ [ف] ند يح، ع ,[ع] نخ، لح ,[ط]لح	نج ♉، لح ♉
Sicily in the West	لو، ,[ط] بو مح، لد ب ,[ر] لو، مح ,[خ] لو، مح [ك] -، لج ,[ف] لر، مح ,[ع]له	
al-Andalus in the West	لب، ,[ط] ي، لح ,[ر] د، بب ,[خ] د، بب [ف] ب، بب ,[ع]لح	ى ♉، لج ♉
The Mediterranean	[ف] لو، ع	

1 Crossed out in the manuscript.

Cities of the fifth clime	Manuscripts	Al-Khwārizmī
The lands of Gog and Magog	نعا، ,[ط] قعا، سجه ,[ر] فما، مخ ,[ج] ما، مح [ك] معا، سجـ مت ,[ع]مح	فا كه، مج له[1]
The lands of the Khāqān	فد، ,[ع] قى، مس ,[ط] قز، مب ,[ج] فى، مس [ك]م سـ كه	
al-Ṭarāz in the lands of the Turks	نو,[ع] قه، م ,[ط] قه ل، م كه ,[أ] فر ن، لط ل [ك]ل، م نو	ق ل، م كد
Isbījāb in Sughd (Sogdia)	صح، ,[ع] صح، م ,[ط] صح، م ,[ج] ا ل، ص له [ك]لط	صج ى، لط ن
Khwārizm in Transoxiana	فه، له لط ,[ع] صا مد ,[ط] قه، مو ,[ج] سح، ر م [ك]	
Ardabīl in Azarbāyjān (Azerbaijan)	عح، ,[ع] عح، م ,[ط] عجـ، م ,[ج] ما د، مب ک [ك]لط	
Khilāṭ in Armenia	سز، لط ,[ج] عد، م ,[أ] سه ♉، لط م [ك] سر سا، له طو ,[ع] سر، لط ,[ط]ل	سد ن، لط ن
Malaṭyā	سا، ,[ج] سر ب، لط ,[أ] سا ♉، لط ♉ [ع] سا، لط ,[ط]لط	سا ♉، لط ♉
Macedonia	[ع] قعد، ر ,[أ][2] مط ♉، م ♉	
Greater Rome of the Romans	,[ط]له لد، ما نر ,[ج] لما، ما ن ,[أ] له ى، ما ن [ع]له، ما	له كه، ما ن
Port-Vendres	[ع] كح، مح ,[ط] كح، مجـ ,[أ] نط ل، مس ى	نط ل، مب ى
The lands of *s-ṭ-a-?-a* in the West	[ك] لح، - ,[ج] له، فح	مه م، لح ♉

1 Al-Khwārizmī lists two cities: *madīnat yājūj* (فاكه، مج له), which seems to correspond to the co-ordinates given in the manuscripts of the epistle, and *madīnat mājūj* (فعا ♉، مه ♉).

2 In MS [أ] the scribe has repeated the toponym Malaṭyā instead of Macedonia.

Cities of the sixth clime	**Manuscripts**	**Al-Khwārizmī**
The lands of Gog	س مع، له ,[ط] فع مب، له ک ,[أ] فر ♉، مس ♉ [ك] ک	
Sīstān in the lands of the Turks	[ط] عا لا، ما كه	
The lands of the Kīmāk Turks	[ك] عا لا، ما له ,[ط] صا فه، مب ي	
The lands of the Khazars	[ك] عح مه، لط م ,[أ] صحـ ♉، مه ♉	صح ♉، مه ♉
Island in the Sea of Jurjān	[ك] عح، ما كه	
Bardhā'a in Azerbaijan	[ك] مه، لط ,[ر] عجـ، لط م ,[أ] عحـ ♉، محـ ♉	عح ♉، مح ♉
Mount Darband	[ك] كو، م ,[غ] ,[ط] مه، مب	
Al-Shiḥr	[ك] كه، مب	
Constantinople	[ك] مط، مر ,[ط] مط نه، م لا ,[أ] مط ن، مه ♉	مط ن، مه ♉
Hargla	[ك] لق، له ,[ط] كر كه، مز له	نج كه، مو له
The lands of the Burjān	[ك] ل، له ,[أ] م ♉، مه ♉	م ♉، مه ♉
The lands of *a-f-y-l-y-w-n* (?)[1]	[ك] سـ مع، م ,[أ] كه مه، مد نه	

1 This toponym seems to be an Arabic transliteration of Albion, see note below.

Cities of the seventh clime	**Manuscript**	**Al-Khwārizmī**
The lands of Gog and Magog	[غ] قعد ك، سجـ لا ,[ط] قعد ك، سجـ لا	
The lands of *al-ʿ-r-h* (?) of the Christians	[ط] نج ♉، مح ♉	
An island in the Western Sea		
Albion (*Alūya*)[1]	[ط] نط م، يط مه	نط م، نط نه

1 *Alūya* is the most common Arabic rendition of the Greek term Albion, cf. D. M. Dunlop, 'The British Isles According to Medieval Arabic Authors', *The Islamic Quarterly*, vol. 4 (1957), p. 1128. Al-Khwārizmī lists the toponym *ʾ-l-w-y* with the following explanation: *wa-hiya madīna kabīra*; cf. *Kitāb Ṣūrat al-arḍ*, p. 34.

Appendix 3

Cities of the seven climes listed in the tables of all the manuscripts

The following tables contain all the cities enumerated in the tables of the twelve manuscripts used for this edition, indicating in which manuscript they occur. Due to the defective script of the copies and the great variation in the rendition of the names, it has not been possible to identify all the toponyms and some of the translations are conjectural. The Arabic text of all the tables can be consulted in the appendix, and where appropriate transliterations of the Arabic names have been provided. The variants of all the unidentified toponyms have been transliterated in the footnotes, and also the variants that are relevant to the study of the textual relationships of the manuscripts.

Major cities of the seven climes	**Manuscripts**
Kashmir in Hind	[أ], [خ], [ر], [ط], [غ], [ف], [ل]
Island of Kark in Hind	[خ], [ر], [ط], [غ], [ف], [ل]
The City of Perfumes in Hind	[أ], [خ], [ر], [ط], [ع], [غ], [ف], [ل]
Hadhramawt in the Yemen	[أ], [خ], [ر], [ط], [ع], [غ], [ف], [ل]
Waghāra in Abyssinia	[أ], [خ], [ر], [ط], [ع], [غ], [ف]
Kūkū in Abyssinia	[أ], [خ], [ر], [ط], [ع], [غ], [ف], [ل][1]
A city on the Western Sea	[أ], [خ], [ر], [غ], [ف]

1 MS [ل] reads *k-y-ṭ-r-s*.

Cities of the first clime	Manuscripts
A city in furthermost China	[أ], [ج], [خ], [ر], [ط], [ع], [غ], [ف], [ك], [ل]
A city in the islands of China	[أ], [ج], [خ], [ر], [ط], [ع], [غ], [ف], [ك], [ل]
The city of *ʾ-s-?-r-?-ā-r* (?) in China	[ج], [خ], [ر], [ط], [ع], [غ], [ف], [ك], [ل][1]
B-ā-y-w-q-ā / *Ḥ-ā-?-y-r-ā* (?) in Hind	[أ], [ج], [خ], [ر], [ط], [ع], [غ], [ف], [ك], [ل][2]
Khaberis in Hind[3]	[ج], [خ], [ر], [ط], [ع], [ك], [ل][4]
M-s-t-ʿ-l-y (?) in Hind	[ج], [خ], [ر], [ط], [ع], [غ], [ف], [ك], [ل][5]
Oman in the land of the Arabs	[ج], [خ], [ر], [ط], [ع], [غ], [ف], [ك], [ل]

1 The manuscripts read: *ʾ-s-ʿ-r-?-ā-r* [ج], *ʾ-s-t-q-r-l-w-y* [خ], *ʾ-s-t-ʿ-r-m-ā-l-w-ṣ-y* [ر], *ʾ-s-y-q-r-?-ā-w-m-y-f-r* [ط], *ʾ-s-?-q-r-ā-w-h-y* [غ], *ʾ-s-f-r-ā-w-h-y* [ف], *ʾ-s-f-r-y-ā-d-w-b-y* [ك], *ʾ-s-w-y-ā-r-w-b-y* [ل], *s-ʿ-w-y-ā-w-h-y* [ع]. Al-Farghānī's rendition of this toponym is *-ā-sh-f-t-y-r-a wa-hiya marqā al-Ṣīn*', cf. *Elementa Astronomica*, p. 35 of the Arabic text, which seems to correspond to al-Khwārizmī's *ʾ-s-f-th-r-ā*, cf. *Kitāb Ṣūrat al-arḍ*, p. 11.

2 The manuscripts read: *b-a-y-w-q-a* [أ], *?-a-?-w-f-a* [ج], *ḥ-a-y-?-w-r-a* [خ], *m-a-?-?-w-w-a* [ر], *?-a-?-?-w-q-a* [ط], *?-a-q-w-m-a* [ع], *b-a-n-b-w-f-a* [غ], *?-a-m-w-q-a* [ف], *?-a-?-?-w-f-a* [ك], *?-a-s-w-?-a* [ل]. This may correspond to al-Khwārizmī's *Kh-ā-n-f-w-ā*, cf. *Kitāb Ṣūrat al-arḍ*, p. 6.

3 Khaberis, the emporium described by Ptolemy, probably corresponds to Kaveripumppattinam.

4 The manuscripts read: *ḥ-ā-b-r-w-n* [ج], *j-ā-b-y-r-w-n* [خ], *ḥ-ā-?-r-w-n* [ر], *ḥ-ā-?-r-w-n* [ط], *ḥ-a-?-r-w-r* [ع], *ḥ-ā-b-r-n* [ك], *ḥ-ā-?-r-w-n* [ل].

5 The manuscripts read: *s-ʿ-l-y* [ج], *s-?-l-y* [خ], s-s-q-l-y [ر], *m-s-t-ʿ-l-y* [ط], *m-s-t-ʿ-l-y* [ع], *m-s-t-ʿ-l-y* [غ], *?-s-q-l-y* [ك], *s-q-l-y* [ل]. This may refer to the city of Masuli, the Maisoleos of Ptolemy.

The city of *al-m-y-d* (?) in Hind[1]	[ج], [خ], [ر], [ط], [ع], [غ], [ف], [ك], [ل][2]
Another city	[ج], [خ], [ر], [ط], [ع], [غ], [ف], [ك], [ل]
Ẓafār in the land of ʿĀd	[ط], [ع], [غ], [ف]
Aden in Yemen	[أ], [ج], [خ], [ر], [ط], [ع], [غ], [ف], [ك], [ل]
Sanʿa in Yemen	[ر],[3] [ط], [غ], [ف]
Dongola in Nubia	[أ], [ج], [خ], [ر], [ط], [ع], [غ], [ف], [ك], [ل][4]
Kusti	[ج], [ر], [ط], [ع], [غ], [ف], [ك], [ل][5]
The kingdom of Abyssinia	[أ], [ج], [خ], [ر], [ط], [ع], [غ], [ف], [ك], [ل]
The Greater Jarma[6]	[ج], [خ], [ر], [ط], [ع], [غ], [ف], [ك], [ل]
The lands of the Berbers	[أ], [خ], [ر], [ط], [ع], [غ], [ك], [ل]
Libya	[ط], [ف]

1 This toponym occurs as *al-Mayd* or *al-Maydh* in al-Khwārizmī and as *al-Mand* in al-Idrīsī, cf. Nallino, 'Al-Ḫuwārizmī e il suo rifarcimento della Geografia di Tolomeo', p. 39.

2 The manuscripts read: *al-m-t-d-a* [ج], *al-m-t-d-a* [خ], *al-m-b-d* [ر], *al-m-n-d* [ط], *al-h-n-d* [ع], *al-m-s-n-d* [غ], *al-m-n-d* [ف], *al-m-?-d* [ك], *al-m-b-d* [ل].

3 Ms. [ر] reads *Yaman al-Yaman.*

4 Ms. [ل] reads *bilād al-Nūba.*

5 The manuscripts read: *k-w-s wa-aʿlā minhu* [ج], *k-w-s wa-a-ʿ-l-y* [ر], *k-w-sh wa-a-ʿ-l-y minhu* [ط], *k-w-s* [ع], *k-w-sh wa-a-ʿ-l-y minhu* [غ], *k-w-sh wa-a-ʿ-l-h min al-nūba* [ف], *k-w-sh wa-a-ʿ-l-y minhu* [ك], *l-w-s wa-a-ʿ-l-h minhu* [ل].

6 Jarma, in Abyssinia, the ancient Aksum.

Cities of the second clime	**Manuscripts**
The city of *ṭ-w-m-ā* (?) in China	[أ], [خ], [ر], [ط], [ع], [غ], [ك], [ل][1]
The city of *ṭ-ʿ-w-l-ā* (?) in China	[أ], [ج], [خ], [ر], [ط], [ع], [غ], [ك], [ل][2]
The city of *ʾ-ṭ-w-ā-s-ā* (?) in China	[ج], [خ], [ط], [ع], [غ], [ك], [ل][3]
Quni in Hind	[ج], [خ], [ر], [ط], [ع], [غ], [ك], [ل]
A city in a mountain	[ج], [خ], [ر], [ط], [ع], [غ], [ك], [ل]
Al-Manṣūra (Brahmanābād) in Sind	[أ], [ج], [خ], [ر], [ط], [ع], [غ], [ك], [ل]
Al-Bīrūn by the sea	[ج], [خ], [ر], [ط], [ع], [غ], [ك], [ل][4]
Daybul	[ج], [ط], [غ],[5] [ك], [ل]
Diyār Bakr[6]	[ج], [ط], [ع],[7] [غ], [ك], [ل][8]
Yamāma in the Ḥijāz	[ج], [خ], [ر],[9] [ط], [ع], [غ], [ك], [ل][10]
Ṭā'if in Yemen	[أ], [ج], [خ], [ر], [ط], [غ], [ك], [ل]

1 The manuscripts read: *ṭ-w-ʿ-m-a* [أ], *ṭ-w-s-a* [خ], *ṭ-w-s-a* [ر], *ṭ-ʿ-w-m-a* [ط], *ṭ-gh-w-l-a* [ع], *ṭ-ʿ-r-?-a* [غ], *ṭ-w-m-a* [ك], *ṭ-w-?-a* [ل].

2 The manuscripts read: *ṭ-ʿ-w-l-a* [أ], *ṭ-w-l-a* [ج], *ṭ-f-w-l-a* [خ], *ṭ-f-w-l-a* [ر], *ṭ-f-r-l-a* [ط] *ṭ-ʿ-r-h* [ع], *ṭ-ʿ-w-l-a* [غ], *ṭ-gh-w-l-a* [ك], *ʾ-ṭ-r-a-t-?-a* [ك], *ṭ-gh-w-l-a* [ل].

3 The manuscripts read: *a-ṭ-w-a-s-a* [ج], *ʾ-ṭ-w-ā-?-ā* [خ], *ʾ-ṭ-w-ā-s-ā* [ط], *ʾ-ṭ-r-s-ā* [ع], *ʾ-ṭ-w-ā-s-ā* [غ], *ʾ-ṭ-w-ā-n-y-ā* [ل]. In al-Khwārizmī's manuscript this toponym occurs as *madīna ʾ-ṭ-r-ā-s-ā*, cf. *Kitāb Ṣūrat al-arḍ*, p. 11.

4 The manuscripts read: *al-s-d-w-r* [ج], *al-b-r-w-r minhu ʿalā al-baḥr* [خ, ر, ك, ل], *al-b-r-w-z minhu ʿalā al-baḥr* [ط], *s-r-w-d min al-Hind* [ع], *al-?-r-w-n minhu ʿalā al-baḥr* [غ]. This toponym may occur both as *al-Bīrūn* and *al-Nīrūn* in Arabic geographical treatises.

5 MS [غ] reads *al-z-b-y-d-y-h.*

6 Lit. 'the dwellings of the tribe of the Bakr' i.e. the region comprising the upper basin of the river Tigris.

7 MS [ع] reads *diyār s -l-y minhu.*

8 MSS [ك] and [ل] read *diyār t-l-y minhu.*

9 MS [ر] reads *Tiḥāma (sic)* instead of *Yamāma.*

10 MS [ل] reads *al-m-ḥ-a-b-h min al-Ḥijāz.*

Mecca in the Tihāma	[أ], [ج], [ط], [ع], [غ], [ك], [ل]
Yathrib, the City of the Prophet	[أ], [ج], [خ], [ر], [ط], [ع], [غ], [ك], [ل]
Akhmīm in Upper Egypt	[أ], [ج], [خ], [ر], [ط], [ع], [غ], [ك], [ل]
Ifrīqiya in the West	[أ], [ج], [خ], [ر], [ط], [غ], [ك], [ل]
Libya	[ج], [خ], [ر], [ط], [غ],[1] [ك], [ل]
Ma'd in the Tihāma	[خ], [ر]

Cities of the third clime	**Manuscripts**
The city of '-*r-a-b-d-a* (?) in China	[ج], [خ], [ر], [ع], [غ], [ف], [ك], [ل][2]
Qandahār in Hind	[أ], [ج], [خ], [ر], [ط], [ع], [غ], [ف], [ك], [ل]
Kabul in Hind	[أ], [ج], [خ], [ر], [ط], [ع], [غ], [ف], [ك], [ل]
Zaranj in Sijistān	[ج], [خ], [ر], [ط], [ع], [غ], [ف], [ك], [ل]
Ṭīz in Makrān	[خ], [ر], [ط], [ع], [غ], [ف], [ك], [ل]
Sirjān in Kirmān	[أ], [خ], [ر], [ط], [ع], [غ], [ف], [ك], [ل]
Shīrāz in Fārs	[أ], [ج], [خ], [ر], [ط], [ع], [غ], [ف], [ك], [ل]
Ahwāz in Khuzistān	[أ], [ج], [خ], [ر], [ط], [ع], [غ], [ف], [ك], [ل]

1 MS [غ] reads *bilād al-Yunān*.

2 The manuscripts read: *'-r-?-d-ā* [ج], *'-d-ā-m-d-y* [خ], *'-dh-ā-b-d-ā* [ر], *'-r-b-d-ā-n* [ع], *'-dh-ā-?-d-ā* [غ], *'-d-ā-b-d-ā* [ف], *'-r-ā-b-d-ā* [ك], *'-r-ā-n-d* [ل].

Basra in Iraq	[أ],[1] [ج], [خ], [ر], [ط], [ع], [غ], [ف], [ك], [ل]
Kufa in Iraq	[أ], [ج], [ر], [ط], [ع], [غ], [ف], [ك]
Damascus in Syria [al-Shām]	[أ], [ج], [خ], [ر], [ط], [ع], [غ], [ف], [ك], [ل]
Jerusalem in Palestine	[أ], [ج], [خ], [ر], [ط], [ع], [غ], [ف], [ك], [ل]
Fusṭāṭ in Egypt [Miṣr]	[أ], [ج], [خ], [ر], [ط], [ع], [غ], [ف], [ك], [ل]
Alexandria	[أ], [ج], [خ], [ر], [ط], [غ], [ف], [ك], [ل]
Qayrawān in the West	[أ], [ج], [خ], [ر], [ط], [غ], [ف], [ك], [ل]
Tangier in North Africa	[ج], [خ], [ر],[2] [ط], [ع], [غ], [ف], [ك], [ل]
K-?-a-n	[ط]

Cities of the fourth clime	**Manuscripts**
The city of *k-a-sh-ʿ-r* in China[3]	[ج], [خ], [ر], [ط], [ع], [ك], [ل][4]
Tibet in the lands of the Turks	[ج], [خ], [ر], [ط], [ع], [غ],[5] [ف], [ك], [ل]
Samarqand in Transoxiana	[أ], [ج], [خ], [ر], [ط], [ع], [غ], [ف], [ك], [ل]

1 MS [أ] reads *al-maqṣūra* instead of *al-Baṣra*.
2 MS [ر] reads *al-m-?-j-a min al-Maghrib*.
3 This toponym may correspond to Ptolemy's Kattighara.
4 The manuscripts read: *k-sh-ʿ-r-a* [ج، خ، ر، ك], *k-sh-gh-r-h* [ط], *k-ā-sh-gh-r* [ع], *k-ā-m-s-w-ā-n-d* [ل].
5 MS [غ] reads *al-d-l-t*.

Balkh in Khurāsān	[أ], [ج], [خ], [ر], [ط], [ع], [غ], [ف], [ك], [ل]
Harāt in Khurāsān	[أ], [خ], [ر], [ط], [ع], [غ], [ف], [ك], [ل]
Marw in Khurāsān	[أ], [ج], [خ], [ر], [ط], [ك]
Nishāpūr in Khurāsān	[أ], [ج], [خ], [ر], [ط], [غ], [ف], [ك], [ل]
Jurjān	[أ], [ج], [خ], [ر], [ط], [ع], [غ], [ف], [ك], [ل]
Āmul in Ṭabaristān	[أ], [ج], [خ], [ر], [ط], [ع], [غ], [ف], [ك], [ل]
Ray in Fārs	[أ], [ج], [خ], [ر], [ط], [غ], [ف], [ك], [ل]
Daylam in Jīlān	[أ], [ج], [خ], [ر], [ط], [ع], [ف], [ك], [ل]
Iṣfahān in Fārs	[أ], [ج], [خ], [ر], [ط], [ع], [ف], [ك], [ل]
Hamadhān in the Land of Māhān	[أ], [ج], [خ], [ر], [ط], [ع], [غ], [ف], [ك], [ل]
Baghdad in Iraq	[أ], [ج], [خ], [ر], [ط], [ع], [غ], [ف], [ك], [ل]
Mosul in the Diyār Rabīʿa	[أ], [ج], [خ], [ر], [ط], [ع], [غ], [ف], [ك], [ل]
Aleppo in Syria	[أ], [خ], [ر], [ط], [ع], [غ], [ف], [ك], [ل]
Cyprus	[خ], [ر], [ط], [ع], [غ], [ف], [ك], [ل]
Anṭākya	[أ], [خ], [ر], [ع], [غ], [ف], [ك], [ل]
Amorium	[أ], [خ], [ر], [ط], [ع], [غ], [ف], [ك], [ل]

Sicily in the West	[خ], [ر], [ط], [ع], [غ], [ف], [ك], [ل]
al-Andalus in the West	[أ], [ج], [خ], [ر], [ط], [ع], [ف], [ك], [ل]
The Mediterranean	[ف]

Cities of the fifth clime	**Manuscripts**
The lands of Gog and Magog	[أ], [ج], [خ], [ر], [ط], [ع], [غ], [ف], [ك], [ل]
The lands of the Khāqān of the Turks	[أ], [ج], [خ], [ر], [ط], [ع], [غ], [ف], [ك], [ل]
Ṭarāz in the lands of the Turks	[خ], [ر], [ط], [ع], [غ], [ف], [ك], [ل]
Isbījāb in Soghdiana	[ج], [خ], [ر], [ط], [ع],[1] [غ], [ف], [ك], [ل]
Khwārizm in Transoxiana	[أ], [ج], [خ], [ر], [ط], [ع], [غ], [ف], [ك], [ل]
Ardabīl in Azerbaijan	[أ], [ج], [خ], [ر], [ط], [ع], [غ], [ف], [ك], [ل]
Khilāṭ in Armenia	[أ], [ج], [خ], [ر], [ط], [ع], [غ], [ف], [ك], [ل]
Malaṭyā in Armenia	[أ], [ج], [خ], [ر], [ط], [ع], [غ], [ف], [ل]
Macedonia	[ج], [خ], [ر], [ط], [ع], [غ], [ف], [ك], [ل]
Greater Rome of the Romans	[أ], [ج], [خ], [ر], [ط], [ع], [غ], [ف], [ك], [ل]
Port-Vendres	[أ], [ج], [خ], [ر], [ع], [غ], [ف], [ك], [ل]

1 MS [ع] reads *a-s-j-a-m*.

The lands of *s-ṭ-a-?-a* (?)in the West	[ج], [خ], [ر], [ط], [ع], [غ], [ك], [ل][1]

Cities of the sixth clime	**Manuscripts**
The lands of Gog	[أ], [خ], [ر], [ط], [ك], [ل]
Sīstān in the lands of the Turks	[ط], [غ], [ف]
The lands of the Kīmāk Turks	[ج], [خ], [ر], [ط], [غ], [ف], [ك], [ل]
The lands of the Khazars	[أ], [ج], [خ], [ر], [غ], [ف], [ك], [ل]
Island in the Sea of Jurjān	[أ], [ج], [خ], [ر], [ط], [غ], [ف], [ك], [ل]
Bardhāʿa in Azerbaijan	[أ], [ج], [خ], [ر], [ط], [غ], [ف], [ك], [ل]
Darband	[أ], [ج], [خ], [ر], [ط], [غ], [ف], [ل]
Al-Shiḥr	[ج], [خ], [ر], [ط], [غ], [ف], [ك], [ل]
Constantinople	[أ], [ج], [خ], [ر], [ط], [ك], [ل]
Hargla	[ج],[2] [خ], [ر], [ط], [غ], [ف], [ك], [ل]
The lands of the Burjān[3]	[أ], [ج], [خ], [ر], [ط], [ف], [ك], [ل]
The lands of *a-f-y-l-y-w-n* (?)[4]	[أ], [ج], [خ], [ر], [ط], [غ], [ف], [ك], [ل][5]
The Western Sea	[ج]

1 The manuscripts read: *bilād s-ṭ-ā-?-ā* [ج], *bilād min al-Maghrib* [خ], [ر], *bilād s-ṭ-ā-dh-ā* [ط], *bilād al-ḥ-r* [ع], *bilād q-s-ṭ-ā-n-ā min al-Maghrib* [غ], *ṭ-r-ā-l-r-m min al-Maghrib* [ك], *bilād s-ṭ-ā-m-ā min al-Maghrib* [ل].

2 MS [ج] reads *m-ʿ-r-w-ṭ*.

3 This term may refer to the Burgundians.

4 This toponym seems to be an Arabic transliteration of Albion, usually rendered as *Alūya*, see n. 38.

5 The manuscripts read: *bilād ʾ-f-y-l-y-w-n* [أ], *bilād ʾ-s-y-l-w-n* [ج], *bilād ʾ-?-l-w-n min al-Maghrib* [خ], *bilād ʾ-m-l-w-n min al-Maghrib* [ر], *bilād ʾ-w-l-y-w-n min al-Maghrib* [ط], *bilād ʾ-q-?-l-w-l min al-Maghrib* [غ], [ف], *bilād ʾ-f-l-y-m-w-n* [ك], *bilād ʾ-h-l-w-n min al-Maghrib* [ل].

Cities of the seventh clime	Manuscript
The lands of Gog and Magog	[أ], [ج], [خ], [ر], [ط], [غ], [ف], [ك], [ل]
The lands of the Toghuzghuz	[ج], [خ], [ر], [غ], [ف], [ك], [ل]
The lands of the Alān Turks	[أ], [ج], [خ], [ر], [غ], [ف], [ك], [ل][1]
The lands of *al-ʿ-r-h* (?) of the Christians	[ج], [خ], [ر], [ط], [غ], [ف], [ك], [ل][2]
An island in the Western Sea	[ج], [خ], [غ], [ف], [ك], [ل]
Alūya[3]	[ط]
An island in the Western Sea	[أ], [ج], [خ], [ر], [ك], [ل]
An island in the Western Sea	[أ], [ج], [خ], [ر], [ك], [ل]
An island in the Western Sea	[أ], [ج], [خ], [ر], [ك], [ل]
W-r-m-y-m or *W-r-y-m* (?)	[ك], [ل]

The first table of manuscript [ع] does not fully correspond to that of the rest of the manuscripts, and includes the following lands and cities: the Island of *k-l-y-h* beyond the Equator, the Island of Kark, a city in the Island of Sapphires, the Island of Perfumes, the Island of *al-?-ṣ-l*, Hadhramawt, *s-w-m-n-ā-t* in China, Waghāra, *al-ṣ-m-a-r-h* in Hind, Kūkū, *?-s-m-h* in Hind, and *ʾ-ʿ-ā-ʾ-b* in Sirindīb.

Manuscript [غ] contains an extra table including the following lands and cities: Khurāsān, Tabrīz, Shīrāz, *B-l-kh-?-s*, *al-s-r-w-a*, Edessa, Khāqān, *Q-dh-h-w-n*.

1 The manuscripts read: *al-ṭāq* [أ], *al-ṭalān* [ج، خ، ر، ك، ل], *al-lān* [غ], [ف].

2 The manuscripts read: *bilād a-b-gh-r min al-Rūm* [ج], *bilād al-gh-r-h min al-Rūm* [خ], [ر], *a-h-r-h* [ط], *bilād al-ʿ-r min al-Rūm* [غ], *bilād al-ʿ-r-h min al-Rūm* [ف], [ل], *bilād r-ʿ-h min al-Rūm* [ك].

3 *Alūya* is the most common rendition of the toponym Albion, the Greek name for Great Britain. Al-Khwārizmī lists the toponym *ʾ-l-w-y* with the following explanation: *wa-hiya madīna kabīra*; cf. *Kitāb Ṣūrat al-arḍ*, p. 34.

Appendix 4

Tables of cities and co-ordinates

The following tables are individual – non critical – editions of each one of the tables with cities and co-ordinates as they appear in the twelve manuscripts, preserving the numerous oddities, the original use of diacritics, misspellings and mistakes. The format used for the edition of these tables is intended to match, as far as possible, that of the manuscripts, including the numerous empty cells and the inaccuracies in the distribution of the data. I have only used *sic* to draw attention to the mistakes in the orthography of common toponyms, to repetitions of names, and to the unusual occurrence of expressions that are not toponyms.

Manuscript *alif* (أ)

Manuscript أ Table 1 Major cities of the seven climes		
العرض	**الطول**	**أسماء المدن**
		قشمير من الهند
د ل		مدينه الطب من الهند
ىد ♉	عا ♉	حضرموت من اليمن
		زغاوه من الحبشة
هـ ♉		كوكو من الحبشة
		مدينه على البحر العربي

Manuscript أ Table 2 Cities of the first clime		
العرض	**الطول**	**أسماء المدن**
		مدينه في أقصى الصين
		مدينه في جزائر الصين
ف ♉	فل ♉	بايوقا من الهند
نح ♉	سه ل	عدن من اليمن
بد ♉	نح م	ديقلة من بلاد النوبة
		مملكة الحبشة
		بلاد البربر

Manuscript أ Table 3 Cities of the second clime		
العرض	**الطول**	**أسماء المدن**
كـ مه	قمب م	طوعما من الصين
كد ل	قح ى	طعولا من الصين

كب ♉	فح ♉	المصوره من السند
كا ♉	سر ب	الطايف من اليمن
كا ك	سر ♉	مكه من التهامه
كه ♉	سه ♉	يثرب مدينة الرسول
كد ك	نه ل	اخميم صعيد مصر
		افريقه من العرب

Manuscript أ Table 4 Cities of the third clime		
العرض	**الطول**	**أسماء المدن**
ل ♉	قى ♉	القندهار من الهند
عح ♉	ق ♉	كابل من الهند
لب ل	قح ♉	السرجان من كرمان
ل ♉	عح ♉	شيراز من فارس
لب ♉	عه ♉	الاهواز من خوزستان
لا ♉	عد ♉	المنصوره (sic) من العراق
لب ♉	سط ♉	الكوفه من العراق

لح ♉	س ♉	دمشق من الشام
لح ♉	نو ♉	ست المقدس من فلسطين
كط ♉	نه ♉	الفسطاط من مصر
لا ♉	نط ♉	الاسكندرية
لب ♉	لا ♉	القيروان من العرب

Manuscript أ Table 5 **Cities of the fourth clime**		
العرض	**الطول**	**أسماء المدن**
م ♉	فح ك	سمرقند وراء النهر
لح م	فح له	بلخ من خراسان
		هرا من خراسان
لح ن		مرو من خراسان
لد ♉	ف مه	نيسابور من خراسان
		جرجان من الجبل
		آمل من طبرستان
له مه	له مه	الريّ
		الديلم من جيلان
لب كه	فو ♉	اصفهان من فارس
لو ♉	عح ♉	همذان من بلاد ماهان
لح كه	ع ♉	بغداد من العراق
لو ل	سح ى	الموصل من ديار رسعه
له ♉	سح ♉	حلب من الشام

لد ى	سا لد	انطاكيه
م ♉	نح ♉	عموريه
		اندلس من المغرب

Manuscript أ Table 6 **Cities of the fifth clime**		
العرض	**الطول**	**أسماء المدن**
		بلاد ياجوج وماجوج
		بلاد خاقان من الترك
لط ل	فر ن	الطراز من الترك
		خوارزم من ما وراء النهر
		اردبيل من اذرسجان
لط م	سه ♉	خلاط من ارمينية
لط ♉	سا ♉	ملطيه من ارمننه
م ♉	مط ♉	ملطيه من ارمينيه[1]
ما ن	له ى	روميه الكبرى من الروم
مب ى	نط ل	هيكل الزهرة

Manuscript أ Table 7 **Cities of the sixth clime**		
العرض	**الطول**	**أسماء المدن**
مب ♉	قر ♉	بلاد ياجوج
مه ♉	صخ ♉	بلاد الجزر

1 The city occurs twice in the manuscript.

		جزيره في بحر جرجان
مح ♉	عح ♉	برذعه من آذرسحان
		باب الابواب
مه ♉	مط ن	قسطنطسه الروم
مه ♉	م ♉	بلاد برجان
مد نه	كه مه	بلاد افلسون

Manuscript أ Table 8 Cities of the seventh clime		
العرض	**الطول**	**أسماء المدن**
		بلاد ياجوج وماجوج
		بلاد الطاق من الترك
		جزىره فى بحر المعرب
		جزىره
		جزىره

Manuscript *jīm* (ج), lacking the first table

Manuscript ج Table 2 Cities of the first clime		
العرض	**الطول**	**أسماء المدن**
		مدينه فى اقصى الصين
		مدينه فى جزاير الصين

		اسعرٮار فى الصين
		ٮاٮوفا من الهند
		خاٮروں من الهند
		سعلى من السند
		عمان فى بلاد العرب
		المتد من السند
		مدينه اخرى على الٮحر منه
		عدن من بلاد اليمن
		بلاد اليمن نفسها
		دٮعله من بلاد النوبه
		كوس واعله منه
		مملكه الحبشه
		حرمى الكبرى
		بلاد البربر

Manuscript ج Table 3 **Cities of the second clime**		
العرض	**الطول**	**أسماء المدن**
		طوما من الصين
		طعولا من الصين
		اطواسا من الصين
		ڡوى من الهند
		مدٮنه فى سع جبل منه
		المنصوره من السند
		الٮدور منه على البحر
		الدنبل منه
		ديار بكر منه

		الحامد (sic) من الحجاز
		طايف من اليمن
		مكة من تهامة
		يثرب مدينه الرسول
		اخميم من صعيد مصر
		افريقيه من المغرب
		بلاد لسواى

Manuscript ج Table 4 Cities of the third clime		
العرض	**الطول**	**أسماء المدن**
لح ل	فس	اربدا من الصين
ل	فى	القندهار من الهند
كح	ن	كابل من بلاد الهند
لب ك	ط به	رويح من السحتان
كط ل	صد م	تىز مكران زمين (sic)
لد د	مح	السرجان من الكرمان
ل	كح	شيراز من فارس
لب	عد	الاهواز من خورستان
لا	عد	البصرة من العراق
لا ر	سطر	الكوفه من العراق
لح	س	دمشق من الشام
لب	ند	بيت المقدس من فلسطين
كله سر	ند م	الفسطاط من مصر
لا	ب	الاسكندريه بطلميوس
لا مه	لا به	القىروان من المغرب
عط لد	با ح	طنجة من المغرب

<table>
<tr><th colspan="5">Manuscript ج Table 5
Cities of the fourth clime</th></tr>
<tr><th>العرض</th><th>الطول</th><th colspan="3">أسماء المدن</th></tr>
<tr><td></td><td></td><td colspan="3">كاشعر من الصين</td></tr>
<tr><td>بح</td><td>بد</td><td colspan="3">اسعب من بلاد الترك</td></tr>
<tr><td>لون</td><td>بك</td><td colspan="3">سمرقند من وراء النهر</td></tr>
<tr><td>لح م</td><td>لح ل</td><td colspan="3">بلخ من حراسان فارس</td></tr>
<tr><td>لح ن</td><td>بد ل</td><td colspan="3">مرو من خراسان</td></tr>
<tr><td>لر</td><td>ر به</td><td colspan="3">جرجان من الحسد (sic)</td></tr>
<tr><td>لح ل</td><td>ں</td><td colspan="3">امل من طبرستان</td></tr>
<tr><td>لر مه</td><td>عو ل</td><td colspan="3">الرى من فارس</td></tr>
<tr><td>له</td><td>عه</td><td colspan="3">الديلم وحىلان</td></tr>
<tr><td>لح ر</td><td>عد</td><td colspan="3">اصفهان من فارس</td></tr>
<tr><td>لد ن</td><td>عد م</td><td colspan="3">همدان من بلاد ماهان</td></tr>
<tr><td><s>لو</s>[1]</td><td>عح</td><td colspan="3">بغداد من العراق</td></tr>
<tr><td>لو</td><td>عح</td><td colspan="3">الموصل من ديار ربيعه</td></tr>
<tr><td>لح</td><td>ع</td><td colspan="3">حلب من الشام</td></tr>
<tr><td>لب ر</td><td>عط</td><td>لد فى</td><td>ساله</td><td>صعله[2]</td></tr>
<tr><td>سح اله ل</td><td>لح عمه</td><td colspan="3">من المغرب محا لح</td></tr>
<tr><td colspan="5">بد مح اندلس له س بو ب</td></tr>
</table>

1 Crossed out in the manuscript.
2 This is the actual arrangement of the cells in the manuscript.

Manuscript ج Table 6 Cities of the fifth clime		
العرض	الطول	أسماء المدن
مح	ما	بلاد ياجوج وماجوج
مب	فى	بلاد خاقان من الترك
ص له	ال	اسحك من السعد
ر م	سح	خوازم (sic) من وراء النهر
مب ك	ماد	الديل من ادرىجان
م	عد	خلاط من ارمىنه
لط	سر ب	ملطيه من ارمىنيه
		ماقادوسه
ما ن	لما	رومىه الكبرى من الروم
		هيكل الزهره من الروم
فح	له	بلاد سطابام

Manuscript ج Table 7 Cities of the sixth clime		
		أسماء المدن
		بلاد ياحوح
		بلاد كىماك
		بلاد الحىدر (sic)
		جزىره فى جرجان
		بردعه ادرىجان
		جبل باب الابواب
		بلاد تخ (sic)
		بلاد سرفطه
		قسطنطىنيه

بلاد بورىحان		
بلاد ابلون		
بحر المغرب		

Manuscript ج Table 8 Cities of the seventh clime		
أسماء المدن	**الطول**	**العرض**
بلاد ياجوج وماجوح		
بلاد التعزعز		
بلاد الطلان من الترك		
بلاد ابغر من الروم		
جربره فى بحر المغرب		
جربره ابضا فىه		
جربره فىه		
جربره ابصا فىه		

Manuscript *khāʾ* (خ)

Manuscript خ Table 1 Major cities of the seven climes		
أسماء المدن	**الطول**	**العرض**
فسمر من الهند	ىو د	بار
جزىرة كوك من الهند	فب	ط
مدينه الطيب من	عب	ه
حضرموت من	عا	ن
رعاه من الحبشه	س	ىن
كركر من الحبشه	ىح	نا
مدينه على البحر الغربي		

Manuscript خ Table 2 **Cities of the first clime**		
أسماء المدن	**الطول**	**العرض**
مدينة في أقصى الصين	فعد	طل
مدينة في جزاير الصين	فعب	لو
استقربالوى من الصين	فسد	ط ط
حانمودا من الهند	فكه	يب
خابيروره من السند	فى	نو به
سملى من الهند	فول	لو
عمان في بلاد العرب	فه	به نط
المدا من الهند	ق	نح
مدنه اخرى على البحر منه	نح	به
عدن من اليمن	سد	نح
السمن اصمين		
دبعله من بلاد النوبه (sic)	نح	به
مملكه الحبشه نوخر	ب	نحو
جر من الكندى	لو	لط
بلاد الترر (sic)	نحو	نط
كوس واعلى من		مقدم (sic)

Manuscript خ Table 3 **Cities of the second clime**		
أسماء المدن	**الطول**	**العرض**
طوسا من الصين	فد	ك
طفولا من الصين	فح	كو ك
اطوابا من الصين	فح	دلد

كا	قد	فوبى من الهند
كو	قما	مدينه
لب	فح	المنصوره من السند
لح ل	فسب	البرور منه على البحر
عا ل	عا مه	التهامة من الحجاز
كو	سه	طائف من اليمن
كو[1]	سه	يثرب مدينه الرسول صلى الله عليه وسلم
		اخميم من صعيد مصر
كو	لح	افريقه من المغرب
كو	مح	بلاد لسوابى
كال	عا مه	الرسله من السند
كا	سر	ممه من تهامه

Manuscript خ Table 4 Cities of the third clime		
العرض	**الطول**	**أسماء المدن**
كح	قسد	ادامدى من الصين
ل	فى	المدهار من الهند
كح	ن	كابل من بلاد الهند
	صد يه	رريح من سحيان
بكطل (sic)	صب م	بير مكران
ا ب د	بح	سيرحان في كرمان
ل	عج	شيراز من فارس
ل	عه	الاهواز من حورستان

1 Co-ordinates repeated in the manuscript.

البصرة من العراق	سط ل	لا ن
دمشق من الشام	س	لح
بيت المقدس من فلسطين	ب	لب
الفسطاط من مصر	نو م	لحط به
الاسكندريه بطلميوس	نـ	لا
القيروان من المغرب	لا نه	لا نه
السحه من المغرب	نا ح	عط لب

Manuscript خ Table 5 Cities of the fourth clime		
أسماء المدن	**الطول**	**العرض**
كاشعر من الصين		
البيت من بلاد الترك	لو	نح
سمرقند مما ورا النهر	لط	لون
بلخ من خراسان	لح ل	لح م
هراه من خراسان	فدك	لو ن
مرو من خراسان	لد نه	لو
نيسابور من خراسان	ف	نح ل
جرجان من الجبل	لو ع	لر نه
امد من طبرستان	عه	ل
الري من خراسان	عد	لح ن
الدىلم من جيلان	عد م	لو ن
اصفهان من فارس	عح	لو
همدان من بلاد ماهان	ع	لح
بغداد من العراق	سط	له ل
الموصل من ديار ربيعه	سح	له ل
حلب من الشام	سا له	لو ى

لو	سر	انطاكيه من الشام
لح	ع	جزيره فوص من بحر المرىم
ع	لو	عمريه من الروم
مح	لو	صقليه من الغرب
ب	د	اندلس من المغرب

Manuscript خ Table 6 **Cities of the fifth clime**		
العرض	**الطول**	**أسماء المدن**
		بلاد ياجوج وماجوج
		بلاد خاقان من الترك
		الطرار من بلاد الترك
		اسبحان من الصعيد
		خوارز من بلاد ورا النهر
		اردسل من ادربجان
		اخلاط من ارمينيه
		ملطيه من ارمينيه
		ىاىار وسه
		رومية الكبرى من الروم
		هيكل الزهرة من المغرب
		بلاد من المغرب

Manuscript خ Table 7 **Cities of the sixth clime**		
		أسماء المدن
		بلاد ياجوج
		بلاد كيماك من الترك

		بلاد الحوز من الترك
		جزيرة في بحر جرجان
		بردعة من ادربجان
		متل باب الابواب
		بلاد سحه
		بلاد هرقط من الروم
		قسطنطينيه من الروم
		بلاد بربحان
		بلاد ابلون من المغرب

Manuscript خ Table 8 Cities of the seventh clime		
العرض	**الطول**	**أسماء المدن**
		بلاد ياجوج وماجوج
		بلاد التعرعر
		بلاد الطلان من الترك
		بلاد الغرة من الروم
		جزيره في بحر المغرب
		حربره
		حربره

Manuscript *rā'* (ر)

Manuscript ر Table 1 Major cities of the seven climes		
الطول	العرض	**أسماء المدن**
باد	بو د	قسمين من الهند
ط	قب	جزيره كرك من الهند

ه	عب	مدينه الطيب من السند
ن	عا	حضرموت من اليمن
ىب	س	رعاو من الحبشه
ىا	نح	كوكو من الحبشة
		مدينه على البحر الغربي

Manuscript ر Table 2 **Cities of the first clime**		
الطول	**العرض**	**أسماء المدن**
مح ل	فعد	مدينه في أقصى الصين
لو	فعب	مدينه في جزاير الصين
ط ط	فعد	اسعرمال وصى من الصين
يب	فكه	مانعو را من الهند
نو ىه	فى	جاسرون من الهند
لد	فدل	سسقلي من السند
مط ىه	فه	عمان في بلاد العربي
ىح	ق	المبد من السند
به	لح	مدينه اخرى على البحر منه
فح	سد	عدن من بلاد اليمن
		اليمن اليمن (sic)
به	ىح	دىعله من بلاد النوبه
		كوس واعلى منه
ىحو	ب	مملكه الحبشه
ل	لد	حرمن الكبرى
ىط	ىحو	بلاد البربر

Manuscript ر Table 3 Cities of the second clime[1]		
أسماء المدن	**الطول**	**العرض**
طوسا من الصين	مو	كر
طفولا من الصين	قمح	كو ل
من نهر قوي من الهند (sic)	فد	كا
مدينه	فما	كو
المنصورة من السند	سح	كب
البرور منه على البحر	فسب	سح ل
التمامة (sic) من الحجاز	عا م	كا ل
طايف من اليمن	سد	كو
يثرب مدينه الرسول	سه	كو
احميم من صعيد مصر		
افريقيه من المغرب	لح	كو
بلاد لسواى	مح	كو
مقدم اطوانا من اليمن	فح	دكد
الدسله من السند	عا مه	مقدم كال
معد من تهامه (sic)	سر	كا

Manuscript ر Table 4 Cities of the third clime		
العرض	**الطول**	**أسماء المدن**
لح ل	فد	اذابدا من الصين
ل	فى	القندهار من الهند

1 The columns of this table have this disposition in the manuscript.

كابل من بلاد الهند	ف	لح
رويح من سجسان	صد يه	
سر مكران	صب م	من كطل (sic)
السيرجان من كرمان	يح	لب د
شيراز من فارس	عح	ل
الاهواز من حورستان	عه	ل
البصرة من العراق	عد	لا
الكوفه من العراق	سط ل	لا ن
دمشق من الشام	س	لح
بيت المقدس من فلسطين	بد	لب
الفسطاط من مصر	نب م	كط يه
الاسكندريه بطلميوس	ن	لا
القيروان من المغرب	لا يه	لا يه
المحه من المغرب	ما ح	عط لر

Manuscript ر Table 5 Cities of the fourth clime		
أسماء المدن	**الطول**	**العرض**
كاشعر من الصين		
النعب من بلاد الترك	مد	لح
سمرقند من وراء النهر	لط	لون
بلخ من خراسان	يح ل	فح م
هراة من خراسان	قدك	لو ن
نيشابور من خراسان	لب يه	لو
مرو من خراسان	ف	مح له
جرجان من الجبل	عرك	لر يه
امل من طبرستان	عه	ل

لح ب	عد	الري من خراسان
لد ن	عد م	الديلم من حىلان
لو	عج	اصفهان من فارس
لح	ع	همدان من ماهان
لب ل	سط	بغداد من العراق
له ل	سح	الموصل من ديار ربيعه
لو ى	سا ل	حلٮ من الشام
لو	س ر	انطاكيه من الشام
لح	ع	حرىره قبرس من بحر الڡرىم
ع	لو	عمرىه من الروم
مح	لو	صڡلىه من الغرب
ىب	د	اندلس من المغرب

Manuscript ر Table 6 **Cities of the fifth clime**		
العرض	**الطول**	**أسماء المدن**
مخ	فما	بلاد ياجوج وماجوج
		بلاد خاقان من الترك
		الطرار من بلاد الترك
		اسىحاٮ من الصعيد
		خوارزم من وراء النهر
		اردبيل من ادرسحان
		اخلاط من ارمينية
		ملطيه من ارمينية
		مافادونىه
		روميه الكبرى من الروم

		هيكل الزهره من المغرب
		بلاد من المغرب

Manuscript ر Table 7 **Cities of the sixth clime**		
العرض	**الطول**	**أسماء المدن**
		بلاد ياجوج
		بلاد كيماك من الترك
		بلاد الحر من الترك
		جزيره في بحر جرجان
		بردعة من ادرسحان
		مىل باب الابواب
		بلاد شحر
		بلاد هرقط من الروم
		قسطىطىه من الروم
		بلاد برىحان
		بلاد املون من المغرب

Manuscript ر Table 8 **Cities of the seventh clime**		
العرض	**الطول**	**أسماء المدن**
		بلاد ياجوج وماجوج
		بلاد التعرعسر
		بلاد الطلان من الترك
		بلاد العزة من الروم
		جزيرة في بحر المغرب

		جزيره
		جزيره

Manuscript *ṭā'* (ط)

Manuscript ط Table 1 Major cities of the seven climes		
العرض	**الطول**	**أسماء المدن**
يا د	قز	قشمير من الهند
طد	قب	جزره كرك من السند
ه	عب	مدينه الطيب من السند
يد	عا	حضرموت من اليمن
ل	س	رعاوه من الحبش
ما	نح	كنطو من الحبش

Manuscript ط Table 2 Cities of the first clime		
العرض	**الطول**	**أسماء المدن**
عح ٮه	ڡعه ل	مدينة في اقصى الصين
يو	قعب	مدينة في جزيرة من الصين
ط ط	سد	اسيقرٮاومى من الصين
يب	قل	ٮاسوقا من الهند
يو ٮه	ڡكه	حاٮرون من الهند
يو	قي	مستعلي من السند
مط ٮه	فد ك	عمان من بلاد العرب
لح	ڡه	المند من السند
ٮه	ٮ	مدينه اخرى على الٮحر منه
نج	سه	ظفار من بلاد عاد

عدن من اليمن	سد	يح
صنعا من اليمن	سد	يد
دنقله من بلاد النوبه	ىح	ىه
كوش واغله منه	ل	نب
مملكة الحبشه	نب	ک
جرمي الكبرى	لب	ک
بلاد البربر	ک	يط
بلاد لىواى	كو	ىط

Manuscript ط Table 3 **Cities of the second clime**		
أسماء المدن	**الطول**	**العرض**
طعوما من الصين	قمد	ک مه
طفرلا من الصين	قمب	كدل
اطراسا من الصين	قم	ا كد
قوي من السند	قد	كد
مدينه سفح جبل منه	قيا	كز
المنصوره من السند	محب	كب
الىروز منه على الىحر	مب	كج ل
الدبيل منه	كب	كدك
ديار بكر منه	مح مه	كد م
اليمامة من الحجاز	عا مه	كا د
الطايف من اليمن	سح	كب
مكه من تهامة	سز	كا
يثرب من المدينه	سه	كه
احميم من صعيد مصر	نه ل	كول

افريقيه من المغرب	له	كو
بلاد الوان	كه	كا

Manuscript ط Table 4 Cities of the third clime		
أسماء المدن	الطول	العرض
القندهار من الهند	قيج	ل ك
كابل من بلاد الهند	قك	كح
زرنج من سجيستان	صد نه	لب ك
تيز مكران	صب م	كط ل
يرجان (sic) من كرمان	فج	لب ز
شيراز من فارس	عح	كط لو
اهواز من خوزستان	عه	لب
البصرة من العراق	عد	لا
الكوفه من العراق	سط ل	لا نه
دمشق من الشام	س	لج
بيت المقدس من فلسطين	نو	لب
الفسطاط من مصر	لز م	كط ىه
اسكندريه بطلميوس	س	لا
القيروان من المغرب	لا مه	لا مه
طنجه من المغرب	ك	له
كهان (sic)	صه	لج

Manuscript ط Table 5 Cities of the fourth clime		
أسماء المدن	الطول	العرض
كاشغره من الصين		
التبت من بلاد الترك	قف	كط لح
سمرقند من ورا النهر	ص	لط
بلخ من خراسان	صد ک	لو ں
هراه من خراسان	فه	لو
مرو من خراسان	فد	لح
نيسابور من خراسان	ف مه	لو
جرجان	ف	لح ں
امل طبرستان	عز ک	لر مه
الري من فارس	عد	لد مد
الديلم وجيلان	عه	لح ى
اصبهان من فارس	عد م	لب كه
همذان من بلاد ماهان	عج	لو
بغداد من العراق	ع	لج
الموصل من ديار ربيعه	سط	له ک
حلب من الشام	سا له	له ي
جزيره قبرس من بحر الروم	سو	لول
عموريه الروم	لح	لح
سقليه من المغرب	لو مح	لد ٮ
اندلس من المغرب	ي	لح

Manuscript ط Table 6 Cities of the fifth clime		
أسماء المدن	الطول	العرض
بلاد ياجوج	قعا	سجه
بلاد خاقان من الترك	قز	مب
الطراز من بلاد الترك	قه ل	م كه
اسبجاب من الصغد	صح	م
خوارزم من ورا النهر	قه	مو
اردبيل من اذرسحان	عج	م
خلاط من ارمينيه	سز	لط ل
ملطيه من ارمينيه	سا	لط
مافادوسه روميه الكبرى من الروم (sic)	له لد	ما نز
هيكل الزهره من المغرب	كح	مج
بلاد سطاذا من المغرب		

Manuscript ط Table 7 Cities of the sixth clime		
أسماء المدن	الطول	العرض
بلاد ياجوج	قع مب	له ک
بلاد سيسان من الترك	عا لا	ما كه
بلاد كيمال من الترك	صا فه	مب ي
جزيرة في بحر جرجان		
برزعه من اذرسجان	عج	لط م
جبل باب الابواب	مه	مب
بلاد شحر		
بلاد هروله من الروم	كر كه	مز له
قسطنطينيه من الروم	مط نه	م لا

		بلاد برجان من الروم
		بلاد اولنون من المغرب

Manuscript ط Table 8 **Cities of the seventh clime**		
العرض	**الطول**	**أسماء المدن**
مح ♉	نح ♉	اهره
نط مه	نط م	الوي
سج لا	قعد ك	مدينه ياجوج

Manuscript *ghayn* (غ)

Manuscript غ Table 1 **Cities of Iran**		
العرض	**الطول**	**أسماء المدن**
٢٣	قد	خراسان
٢٥	كح ٢	توريز (sic)
٢٣	قا ٩	شيراز
٢١	عح ٣	بلخس
٢٧	ص ١	الروا
٢١	كق ٨	الرها
٢٦	مط	خاقان
٢٩	عح	قذمون

Manuscript غ Table 2 **Major cities of the seven climes**		
العرض	**الطول**	**أسماء المدن**
ح	فكر	قشمير من الهند
ط	ف عب	جزيرة كرك من السند

مدينه الطيب من السند	عب	ه
حضرموت من اليمن	عا	ب
رغاوه من الحبشه	لح	يا
كركور من الحبشه		
مدينه على البحر الغربي		

Manuscript غ Table 3 **Cities of the first clime**		
أسماء المدن	**الطول**	**العرض**
مدينة اقصى الصين	يعو ل	يح مه
مدينه في جزيرة من الصين	سف	لط
استقريا وهى من الصين	فكر يا	لز يه
بانبوفا من الهند	قكه	لو
مسعلى من السند	ح	لط مه
عمان في بلاد الغرب	فد ل	ل ح
المسند من السند	فه	له
مدينه أخرى على البحر منه	و	
طغار من بلاد عاد	سمر	يه
عدن من اليمن	لح	يه
صنعا من اليمن	يه	مب
دمعايد من بلاد النوبه	مب	ب
ممالك الحبشة	فد	ك
حرمى الكبدى	كو	ك
بلاد برابر		يط

Manuscript غ Table 4 Cities of the second clime		
العرض	**الطول**	**أسماء المدن**
		طعربا من الصين
		طعولا من الصين
		اطواسا من الصين
		قون من الهند
		مدينه سفل جبل منه
كب ♉	فح ♉	المنصوره من السند
كو ل	قب ک	السرون منه على البحر
		الزبيديه
		ديار بكر منه
كا ل	فا به	اليمامه من الحجاز
كا م	عر	طائف
		مكه من تهامه
		يثرب مدينه رسول الله
		اخميم من صعيد مصر
		افريقيه من الغرب
		بلاد اليونان

Manuscript غ Table 5 Cities of the third clime		
العرض	**الطول**	**أسماء المدن**
		اذاندا من الصين
		القندهار من الهند
		كابل من الهند
		اولخ من سخستان

تيز من مكران		
السعرجان من كران		
اشيراز (sic) من فارس		
الاهواز من حرستان		
البصره من العراق		
الكوفه من العراق		
دمشق من الشام		
بيت المقدس		
الفسطاط من مصر		
الاسكندريه بطلميوس		
القيروان من المغرب		
طنجه من المغرب		

Manuscript غ Table 6 Cities of the fourth clime		
أسماء المدن	الطول	العرض
الدلب من بلاد الترك		
سمرقند من وراء النهر		
بلخ من خراسان		
هواه (sic) من خراسان		
نيسابور من خراسان		
جرجان		
امل من طبرستان		
الري من فارس		
همدان من بلاد ماهان		
بغداد من العراق		

		الموصل من ديار ربيعه
		حلب من الشام
		انطاكيه من الشام
		جزيرة قبوس فى بحر الروم
		عمريه من الروم
		سقليه من المغرب

Manuscript غ Table 7
Cities of the fifth clime

العرض	الطول	أسماء المدن
		بلاد ياجوج
		بلاد خاقان من الترك
		الطراز من بلاد الترك
		اسبجان من الترك
		خوارزم من وراء النهر
		اردبى من ادرىىجان
		ملطيه من ارمينيه
		احلاط من ارمينيه
		مَاقاذونيه
		رومية الكبرى من الروم
		هيكل الزهرة من المغرب
		بلاد نطابا من المغرب

Manuscript غ Table 8 Cities of the sixth clime		
العرض	**الطول**	**أسماء المدن**
		بلاد سىسان من الترك
		بلاد كيماك من الترك
		بلاد خزر من الترك
		جزيرة في بحر جرجان
		بردعه من ادربىجان
		جبل باب الابواب
		بلاد هرقله من الروم
		بلاد اقىلول من المغرب
		بلاد سجير

Manuscript غ Table 9 Cities of the seventh clime		
العرض	**الطول**	**أسماء المدن**
		بلاد ياجوج
		بلاد الان من الترك
		بلاد التغرغر من الترك
		بلاد العس من الروم
		جزيرة في بحر الغرب

Manuscript *fāʾ* (ف), lacking the table with the cities of the second clime

Manuscript ف Table 1 **Major cities of the seven climes**		
أسماء المدن التي في الأقاليم السبعة	**الطول**	**العرض**
قشمير من الهند	فلر	فلر
جزيرة كرك من السند	قب	ط
مدينه الطيب من السند	عب	ه
كنطو (sic) من الحبشة	عا	ب
حضرموت من اليمن	لح	يا
وعاوه من الحبشه		
كوكوا من الحبشة		
مدينه على البحر العوَا		

Manuscript ف Table 2 **Cities of the first clime**		
أسماء المدن التي في الإقليم الأوّل وهو لزحل	**الطول**	**العرض**
مدينه في أقصى اليمن	فعه ل	لح يه
اسفربا وهى من مرقا الصين	سد	ط ط
خانىرون من الهند	فكه	نو يه
عمان في بلاد العرب	فد ك	مط يه
مدينه اخرى على البحر منه	ب	يه
عدن من اليمن	سد	لح
ريقله في بلاد النوبه	لح	يه
المملكه من الحبشه	ب	ک

ٮط	ك	بلاد بربر
ٮو	فعٮ	مدينه في جزيره الصين
ٮ	فل	ٮاموقا من الهند
ٮو	فى	مستعلى من السند
لح	له	المند من السند
لح	سه	ظفا (sic) من بلاد عاد
ٮد	سد	صنعا من اليمن
ٮ	ل	كوش واعله من النوبه
ك	لٮ	حومى الكبرى
ٮط	كو	بلاد ليواى

Manuscript ف Table 3 **Cities of the third clime**		
العرض	**الطول**	**أسماء المدن التي في الإقليم الثالث**
عح ل	فر	اداٮدا من الصين
ل	فى	القندهار من الهند
كح	ں	كابل من بلاد الهند
ٮه	مد	اوٮح (sic) من سحتان
لد د	مح	ٮتر من مكران
لٮ د	مح	الشيرجاں من كرمان
ل	عح	السرار من فارس
لد	عه	الاهواز من خوزستان
لا	عد	البصره من العراق
لا ٮ	ٮط ل	الكوفة من العراق
لح	س	دمشق من الشام
لد	ٮر	بيت المقدس من فلسطين
كط ٮه	ٮد م	الفسطاط من مصر

الاسكندريه بطلموس	ىد	لا
القيروان من المغرب	لا يه	لا يه
الطنجه من المغرب	ما ح	عط لز

Manuscript ف Table 4 Cities of the fourth clime		
أسماء المدن التي في الإقليم الرابع	**الطول**	**العرض**
التبت من بلاد الترك	ىد	ىح
سمرقند من ما ورا النهر	ىط	لون
بلخ من خراسان	ىح ل	لح م
هراه من خراسان	ا مه	لو
نيسابور من خراسان	و	لح ل
حرحان من ماريددان	عد د	لز يه
آمل من طىرستان	عه	ل
الرى من فارس	عد	لح ر
الديلم وحيلان	عد م	لون
اصفهان من فارس	عح	لو
همذان من بلاد ماهان	ع	لح
بغداد من العراق	سط	لد ل
الموصل من ديار ربيعه	سح	له ل
حلب من الشام	سا له	له ى
جزيره قبرس فى بحر الروم	ع	لح
عمريه من الروم	ىد ىح	ع
سقليه من المغرب	لز	مح
اندلس من المغرب	ب	ب
انطاكيه من الشام	مر	لو
ىحر الروم	لو	ع

Manuscript ف Table 5
Cities of the fifth clime

العرض	الطول	أسماء المدن التي في الإقليم الخامس
		بلاد ىاجوح
		بلاد خافان من الترك
		الطراز من الترك
		اسىحان من السغد
		خواررم من ورآ النهر
		اردسو من اذرىىحان
		ملطيه من ارمسه
		اخلاط من ارمىنه
		ماقاذونه
		روميه الكبرى من الروم
		هيكل الزهره من المغرب

Manuscript ف Table 6
Cities of the sixth clime

العرض	الطول	أسماء المدن التي في الإقليم السادس
		بلاد سىسان من الترك
		بلاد كيماك من الترك
		بلاد خزر من الترك
		جزيره في بحر حرحان
		ىردع (sic) من اذرىيجان
		جبل باب الابواب
		بلاد هرقه من الروم

		بلاد ٮرحان من الروم
		بلاد افيول من المغرب
		بلاد سحر

Manuscript ف Table 7 **Cities of the seventh clime**		
العرض	**الطول**	**أسماء المدن التي في الإقليم السابع**
		بلاد ياجوح
		بلاد آلّان من الترك
		بلاد التغزغز من الترك
		بلاد العره من الروم
		حزٮره في بحر المغرب

Manuscript *kāf* (ك)[1]

Manuscript ك Table 1 **Cities of the first clime**					
الطول	**العرض**	**أسماء المدن**	**العرض**	**الطول**	**أسماء المدن**
		مدينه فى اقصى الصين	ط ط	سد	مدىنه فى اقصى الصين
		مدينه في جزىره الصين	ب	فل	مدىنه فى حزىره الصين
		اسمرىادوبى فى الصين	ىو ىه	فكه	اسفريادوي في الصين
		ىاسوفا من الهند	ىو	في	ىاسوفا من الهند
		حاىرون من الهند	مط يه	فد ل	حاهرون من الهند
		ىسقلى من السند	ىج	مه	ىسقلى من السند
		عمان من بلاد العرب	ىح	سد سد	عمان من بلاد العرب
		المَبَد من السند	ىد	سد	المىد من السند
		مدينة اخرى على البحر ميلا	ىه	ىح	مدينه اخرى على البحر ميلا (sic)
		عدن من اليمن	ىب	لك	عدَن من بلاد اليمن
		زنقله من بلاد نوبه	ک	ىت	زغقله من بلاد نوبه
		كروس وواعله منه	ک	لب	كوس وواعله منه

1 This format reproduces the tables as they appear in the manuscript. The names of the cities are repeated in two different columns with occasional variations. The number of cells in the columns is also extremely variable; sometimes the number of cells in the columns for longitude and latitude is different, and the number of cells with the co-ordinates does not always correspond to the number of cities, it being impossible to establish a relationship between places and co-ordinates.

		مملكه الحبشة	ىط	ك	مملكه الحبشة
		حسمى الكبرى	ىط	كو	حسمى الكبرى
		بلاد البربر			

Manuscript ك Table 2 Cities of the second clime					
الطول	**العرض**	**أسماء المدن**	**العرض**	**الطول**	**أسماء المدن**
		طوما من الصين	ك	فمد	طوما من الصين
		طغولا من الصين	مه	ىح	طغولا من الصين
		اطواتيا من الصين	كد	فد	اطراتبا من الصين
		قوي من الهند	له	فا ىح	قوى من الهند
		مدنه فى سفح جبل منه	ا ك	نب لب	المنصوره من السند
		المنصوره من الهند	كد	ىه ىح	مدينة فى سفح جبل منه
		البرور منه على البحر	كر لب	مه عا	الذبيل منه
		الذَبيل منه	لح ل	سح	السرور منه على البحر
		ديار تلى منه	له ك	سر	اليمامه من الحجاز
		اليمامة من الحجاز	كد م	سه	دبار تلى منه
		الطائف من اليمن	كا م	ىه	مكة من التهامه
		مكه من التهامة	لب	ل	الطائف من اليمن
		يثرب مدينة الرسول	كا	له	حميم (sic) من صعيد مصر

		حميم (sic) من صعيد مصر	له	ل	يثرب مدينة الرسول
		افريقية من المغرب	لو ل	ك	بلاد السوابى
		بلاد السوابي	لو ك	ب	افريقية من المغرب

Manuscript ك Table 3 Cities of the third clime					
الطول	**العرض**	**أسماء المدن**	**العرض**	**الطول**	**أسماء المدن**
		اربدا من الصين	لك لج	فخ فك	اربدا من الصين
		القدهار من الهند	ك لد	صد به	القدهار من الهند
		كابل من بلاد الهند	لط ب	صب	كابل من بلاد الهند
		روبح من سجستان	لد ل	نح ع	الروبح من سجستان
		بيز من مكران	لط لو	عه م	تيز من مكران
		السرحان من كرمان	لد لا	[. . .]	السرحان من كرمان
		سيراز من فارس	لب	س	شيراز من فارس
		الاهواز من خوزستان	لح	بو	الاهواز من خوزستان
		البصرة من العراق	لب	لر م	البصرة من العراق
		دمشق من الشام	كط	س	الكوفه من العراق
		بيت المقدس من فلسطين	به لا	به	دمشق من الشام

		المسطاط من مصر الاسكندريه بطلموس	مه لا	لا به	بيت المقدس من فلسطين
		القيروان من المغرب	له	ك	الفسطاط من مصر اسكندريه بطلموس
		طنجه من المغرب	لح		القيروان من المغرب
					طنجة من المغرب

Manuscript ك Table 4 **Cities of the fourth clime**					
الطول	**العرض**	**أسماء المدن**	**العرض**	**الطول**	**أسماء المدن**
		كاشعر من بلاد الصين	كط	قف ص	كاشعر من بلاد الصين
		بيت من بلاد الترك	لط	صد ك	التبت من بلاد الترك
		سمرقند من ما وراء النهر	لور	فد مه	سمرقند من ما ورا النهر
		بلخ من خرسان	نو	ومه	سمرقند من ما وراء النهر[1]
		هراه من خراسان	لج	و	هراه من خراسان
		مرو من خراسان	لو	ك عد	مرو من خراسان
		نيسابور من خراسان	لح لو	عه	نيسابور من خراسان

1 Repeated in the manuscript.

		جرجان من الجبل	لر مه	عد م	جرجان من الجبل
		امل من طرسان	لر مه	عح لو	امل من طيرستان (sic)
		الريّ من فارس	لح ي	سط	الري من فارس
		الديلم وجبلان	كه لب	سا	الديلم وجبلان
		إصفهان من فارس	عج	له	اصفهان من فارس
		همدان من بلاد ماهان	له ك	سو	همدان من بلاد ماهان
		بغداد من العراق	لج	ىح	بغداد من العراق
		الموصل من ديار ربىعه	لو ل	ى	الموصل من ديار ربيعة
		حلب من الشام	ىر ى	عد	حلب من الشام
		انطاكيه من السام حريره قبرس من الروم	فى د	ك	انطاكيه من الشام
		عمورية الروم	لح		جزيرة قبرس فى بحر الروم
		صفالىة (sic) من المغرب	لج		عمورية الروم صقالىه (sic) من المغرب
		اندلس من المغرب			اندلس من المغرب

Manuscript ك Table 5 Cities of the fifth clime					
الطول	**العرض**	**أسماء المدن**	**العرض**	**الطول**	**أسماء المدن**
		بلاد ياجوج وماجوج	سج مت	معا	بلاد ياجوج
		بلاد خاقان من الترك	م ىـ كه	فد	بلاد خاقان من الترك
		الظراز من بلاد الترك	م ىو	ىو ل	الطرار من بلاد الترك
		اسبيحاب من السعد	لط	صح	اسبحاب من السعد
		حوارزم من ورا النهر	له لط	فىه	حوارزم من ورا النهر
		ارذيل من ادرىىحان	لط	عح	ارذيل من اذربىجان
		اخلاط من ارمينيه	له طو	سر سا	احلاط (sic) من ارمينيه
كا ما ىلاد مرا لن[1]	هـ طرالرم من المعرب	ملطيه من ارمنه ماقادوسه	فج	له	رومة الكبرى من الروم هيكل الزهرة من المغرب
		رومية الكبرى من الروم		لج	بلاد سطاما المغرب

1 This is the actual arrangement and text in the manuscript.

Manuscript ك Table 6 **Cities of the sixth clime**					
أسماء المدن	**الطول**	**العرض**	**أسماء المدن**	**العرض**	**الطول**
بلاد ياجوج وماجوج	ب مع	له ك	بلاد ياجوج وماجوج		
بلاد كيماك من الترك	عا لا	ما له	بلاد كيمان من الترك		
بلاد الخزر من الترك	عح مه	لط م	بلاد الخزر من الترك		
جزيره في بحر جرجان	عح	ما كه	ذكر اسما المدن (sic)		
بردعه من اذرسجان	مه	لط	بلاد ياجوج وماجوج		
جبل باب الابواب	كو	م	بلاد كـمان من الترك		
بلاد سحد	كه	مب	بلاد الحرر من الترك حربره من بحر حرجان		
بلاد هرقلة الروم	لق	له	بردعه من اذرسجان جبل باب الابواب		
قسطنطينية من الروم	مط	مر	بلاد سجر		
بلاد رعه (sic) من الروم	ل	له	بلاد هرفطه من الروم قسطنطينة من الروم		

		بلاد بريحان	م	سـ مع	بلاد افليمون من المعرب
		بلاد املغون من المعرب	لا	عا لا	الاقليم السابع (sic)

Manuscript ك Table 7 Cities of the seventh clime		
		أسماء المدن
		بلاد ياجوج وماجوج
		بلاد النغزغر
		بلاد الطلان من الترك
		بلاد العوه من الروم
		جزيرة في بحر المغرب
		جزيرة
		جزيرة
		جزيرة
		جزيرة
		ورمسم (sic)

Manuscript *lām* (ل)

Manuscript ل Table 1
Major cities of the seven climes[1]

العرض	**الطول**	**أسماء المدن**
يا	قكز	قشمير من الهند
ط	قب	جزيرة كرك من السند

1 Not in tabular form in the manuscript.

جزيرة الطيب من السند	عبه	
كيطورس (sic) من الحبش	نح	يا

Manuscript ل Table 2 Cities of the first clime		
أسماء المدن	**الطول**	**العرض**
مدينة فى اقصى الصين		
مدينة في جزاير الصين		
اسوياروىى في الصين		
باسوبا من الهند		
حابرون من الهند		
سقلى من الهند		
عمان في بلاد الغربى		
المبد من السند		
مدينة لوى على البحر ميلا (sic)		
عدن من اليمن		
بلاد النوبه		
لوس واعله منه		
المملكه من الحبشه		
حربره الكندى		
بلاد البربر		

Manuscript ل Table 3 Cities of the second clime		
المدن	**الطول**	**العرض**
طوبا من الصين		
طغولا من الصين		

اطوانيا من الصين		
قدى من الهند		
مدينه في سبع جبل منه		
المنصوره من الهند		
البرور منه على البحر		
الدبيل منه		
ديار بلى منه		
والمجابه (sic) من الحجاز		
طائف من اليمن		
مكه من نهاية (sic)		
يثرب مدينه الرسول عليه السلم		
حميم (sic) من صعىد مصر		
افرنقىه من المغرب		
بلاد ليوابى		

Manuscript ل Table 4 Cities of the third clime		
أسماء المدن	**الطول**	**العرض**
اراند من الصين		
القندهار من الهند		
كابل من بلاد الهند		
روبح من سجستان		
بسر مكران		
السرحان من كرمان		
شيراز من فارس		
الاهواز من خوزستان		
البصرة من العراق		

دمشق من الشام		
بيت المقدس من فلسطين		
الفسطاط من مصر		
الاسكندريه بطلميوس		
الولن من المغرب		
طنجه من المغرب		

Manuscript ل Table 5 Cities of the fourth clime		
أسماء المدن	**الطول**	**العرض**
كامسواند من الصين		
السب من بلاد الترك		
سمرقند من ما وراء النهر		
بلخ من خراسان		
هراة من خراسان		
نيسابور من خراسان		
جرجان من الجبل		
آمل من طبرستان		
الرى		
الدىلم وجبلان		
اصفهان من فارس		
همذان من بلاد ماهان		
بغداد من العراق		
الموصل من ديار ربيعه		
حلب من الشام		
انطاكيه من الشام		
جزىرة فبرس من الروم		

		عمريه الروم
		سقاليه من المغرب
		اندلس من المغرب

Manuscript ل Table 6 Cities of the fifth clime		
العرض	الطول	أسماء المدن
		بلاد ياجوج وماجوج
		بلاد خاقان من الترك
		الطراز من بلاد الترك
		اسسجاب من السعد
		خوازرم من وراء النهر
		اردبيل من اذربيجان
		اخلاط من ارمينية
		ملطية من ارمينية
		ماقادونيه
		رومية الكبرى من الروم
		هيكل الزهرة من المغرب
		بلاد سطاما من المغرب

Manuscript ل Table 7 Cities of the sixth clime		
العرض	الطول	أسماء المدن
		بلاد ياجوج
		بلاد كيمان من الترك
		بلاد الحرر من الترك
		جزيره من بحر جرجان

		براغة (sic) من ادربيجان
		جبل باب الابواب
		بلاد سحر
		بلاد هرقطه من الروم
		قسطنطينيه من الروم
		بلاد برىجان
		بلاد اهلون من المغرب

Manuscript ل Table 8 Cities of the seventh clime		
العرض	**الطول**	**أسماء المدن الّتي في الإقليم السابع**
		بلاد ياجوج وماجوج
		بلاد البغرغر
		بلاد الطلان من الترك
		بلاد العره من الروم
		حزىره في بحر المغرب
		حزىره
		جزىره
		جزىره
		وريم (sic)

Manuscript *nūn* (ن) containing one table only

Manuscript ن Table 1 Major cities of the seven climes		
العرض	**الطول**	**أسماء المدن**
فلر	فلو	قشمير من الهند
ط	عب	جزيره كرك من السند

ه	عب	مدينه الطيب من السند
ب	عا	حضرموت من اليمن
با	لح	وغاوه (sic) من الحبشه
		كركوا (sic) من الحبشه
		مدينه على بحر الغرب

Manuscript *ʿayn* (ع) containing one table only, and lists of the cities in running text

عرض	طول	أسماء المدن	عرض	طول	
ط	قب ك	جزيرة كوك	فى	قع د	جزيرة كلـه من ورا خطّ الاستواء
ه	عب	الطب مدينة من السّند	د س	قع و	مدينة في جزيرة الياقوت
ب	عا	حضرموت اليمن	ر ل	لح ل (في الهامش: صح ل)	جزيرة اليصل
با	س	وفافمس (sic) الحبشة	د	سقط (sic)	سومنات من الصين
با	ع	ليلر مح (sic) من الحبشة	د	قمه	الصماره (sic) من الهند
با	لح	كركوه (sic) من الحبشة	د	س	قسمة من الهند
			ه ح	وط (في الهامش: ذ ح)	اعاب من سرنديب

في الهامش: زغاوه من الحيشه طول س عرض با

Cities of the first clime:

مدينة في أقصى الصين قعه لح مدينة في جزيرة من الصين قم بر ياقوما من الهند قد م حايرور من الهند قكه يو سعويا وهي موما الصين بدح يط مستعلي من السند مر يو عمان من بلاد العرب قد بط مدينة أخرى على البحر فد يه الهند (sic) من السند فد لح طغار من بلاد عاد سد ه عدن من اليمن سد ح زبيلة (sic) من بلاد النوبة لح يو كوس د م مملكة الحبش مد يط حرمي الكبرى لو كه بلاد بربر كو بط

Cities of the second clime:

طغولا من الصين لح كد طعره من الصين قد كر المنصور من السند صح كد برور (sic) من الهند صد كح أطراسا من الصين قم ل الديبل منه صب كد فوى من الهند فد كد ديار سلي منه (sic)مح كد اليمامة من الحجاز عا كا مدينة في سفح جبل ما كو مكّة من تهامة س كا يثرب مدينة الرسول سه كد حمير (sic) من صعيد مصر يه كو

Cities of the third clime:

اربدان من الصين قس لح القندهار من الهند فى ل كابل من الهند ق كح درنج من سجستان سد كد تيز مكران صب كظ سيرجان من كرمان قح لب شيراز من فارس عخ لب أهواز من حوزستان عو لب البصرة من العراق عد لا الكوفة من العراق سط لا دمشق من الشام س لج بيت المقدس من فلسطين يو لب فسطاط من مصر يب لا لا طنجة من المغرب يا كط

Cities of the fourth clime:

كاشغر من بلاد الصين تبت من بلاد الترك فد لح سمرقند ممّا وراء النهر قد لو بلخ من خراسان قد كد هراة من خراسان قد كد الري عه له جرجان ف لح آمل من طبرستان عو لو إصفهان عد لد الديلم والجيلان عه لح همذان كح لو بغداد ع لح حلب سح له الموصل سط له انطاكية سل له عمورية من الروم نح لح سقلية من الروم لو له جزيرة بحر قبروس في بحر الروم س لو أندلس من المغرب لب لح

Cities of the fifth clime:

بلاد ياجوج ىعا مح بلاد خاقان قي مب الطّراز من بلاد الترك قه م اسحام (sic) من الصغد صح م خوارزم صا مد اردبيل من اذربيجان عح م حلاد (sic) من ارمينية سر لط ملطية سا لط ماقادونية من الروم قعد ر رومية الكبرى له ما هيكل الزهرة من المغرب كح مح بلاد قسطانا من المغرب [. . .]

Bibliography

Ikhwān al-Ṣafāʾ

Rasāʾil Ikhwān al-Ṣafāʾ, ed. Buṭrus al-Bustānī. 4 vols. Beirut: Dār Ṣādir and Dār Bayrūt, 1377/1957, 2004.

Kitāb Ikhwān al-Ṣafāʾ wa-Khullān al-Wafāʾ li'l-imam al-humām quṭb al-aqṭāb mawlānā Aḥmad ibn ʿAbd Allāh, ed. Wilāyat Ḥusayn. 4 vols. Bombay: Maṭbaʿat Nukhbat al-Akhbār, 1305–1306/1887–1889.

Rasāʾil Ikhwān al-Ṣafāʾ wa-Khullān al-Wafāʾ, ed. Khayr al-Dīn al-Ziriklī. 4 vols. Cairo: al-Maṭbaʿa al-ʿArabiyya bi-Miṣr al-Tijāriyya al-Kubrā. 1346/1928.

NB The full reference to specific volumes of the current series of editions of the Epistles of the Brethren of Purity is given under 'Primary and Secondary Sources', under the names of the contributors. For Epistles 1–2 see Nader El-Bizri, *On Arithmetic and Geometry*; for Epistle 5 see Owen Wright, *On Music*; for Epistles 10–14 see Carmela Baffioni, *On Logic*; for Epistle 52a see Godefroid de Cattalaÿ and Bruno Haflants, *On Magic I*.

Primary and Secondary Sources

Abdukhalimov, Bahrom. 'Aḥmad al-Farghānī and his *Compendium of Astronomy*', *Journal of Islamic Studies*, vol. 12, 2 (1999), pp. 142–58.

Adamson, Peter, *Al-Kindī*. Oxford: Oxford University Press, 2007.

Ahmad, S. Maqbul. 'Djughrāfiyā', in *EI2*, vol. 2, pp. 575–87.

—— 'Kharīṭa', in *EI2*, vol. 4, pp. 1077–1083.

Arberry, A. J. *The Koran Interpreted: A Translation*. London: Allen & Unwin, 1955.

Baffioni, Carmela. 'Il "quarto clima" nell'Epistola sulla Geographia degli Iḫwān al-Ṣafā'', in *Oriente Occidente. Scritti in memoria di Vittorina Langella*, ed. F. Bencardino. Naples: Instituto Universitario Orientale, 1993, pp. 45–60.

——(ed.) *On Logic. An Arabic Critical Edition and English Translation of Epistles 10–14*. Oxford: Oxford University Press–The Institute of Ismaili Studies, 2010.

Bausani, Alessandro. *L'Enciclopedia dei Fratelli della Purita*. Naples: Instituto Universitario Orientale, 1978.

Becker, C. H. 'Baḥr al-Qulzum', in *EI2*, vol. 1, pp. 931–33.

Becker, C. H. and D. M. Dunlop. 'Baḥr al-Zandj', in *EI2*, vol. 1, pp. 937–38.

van Bladel, Kevin. *The Arabic Hermes: From Pagan Sage to Prophet of Science*. Oxford: OUP, 2009.

Bosworth, C. E. 'Kimäk', in *EI2*, vol. 5, pp. 107–108.

de Callataÿ, Godefroid. *Ikhwān al-Ṣafā'. A Brotherhood of Idealists on the Fringe of Islam*. Oxford: Oneworld, 2005.

de Callataÿ, Godefroid, and Bruno Halflants (ed.) *On Magic. An Arabic Critical Edition and English Translation of Epistle 52a*. Oxford: Oxford University Press–The Institute of Ismaili Studies, 2011.

—— 'The Classification of Knowledge in the *Rasā'il*', in *The Ikhwān al-Ṣafā' and their* Rasā'il. *An Introduction*, ed. Nader El-Bizri. Oxford: Oxford University Press–The Institute of Ismaili Studies, 2008, pp. 58–82.

Canard, M., and Cl. Cahen. 'Diyār Bakr', in *EI2*, vol. 2, pp. 343–47.

Casson, Lionel. *The Periplus Maris Erythraei*. Princeton: Princeton University Press, 1989.

Crone, Patricia. *Nativist Prophets of Early Islamic Iran. Rural Revolt and Local Zoroastrianism*. Cambridge: Cambridge University Press, 2012.

Dieterici, Friedrich. *Die Propaedeutik der Araber im zehnten Jahrhundert*. Hildesheim: G. Olms, 1969.

Doerfer, G. 'Khaladj', in *EI2*, vol. 4, pp. 917–18.

Dunlop, D. M. 'The British Isles According to Medieval Arabic Authors', *The Islamic Quarterly*, vol. 4 (1957), pp. 11–28.

—— 'al-Baḥr al-Muḥīṭ', in *EI2*, vol. 1, p. 934.

—— 'Baḥr al-Rūm', in *EI2*, vol. 1, pp. 934–36.

El-Bizri, Nader (ed.) *The Ikhwān al-Ṣafā' and their* Rasā'il. *An Introduction*. Oxford: Oxford University Press–The Institute of Ismaili Studies, 2008.

—— (ed.) *On Arithmetic and Geometry, An Arabic Critical Edition and English Translation of Epistles 1–2*. Oxford: Oxford University Press–The Institute of Ismaili Studies, 2012.

—— 'Epistolary Prolegomena: On Arithmetic and Geometry', in *The Ikhwān al-Ṣafā' and their* Rasā'il: *An Introduction*, ed. N. El-Bizri. Oxford: Oxford University Press–The Institute of Ismaili Studies, 2008, pp. 180–213.

d'Encausse, C. 'Derbend', in *EI2*, vol. 2, p. 206.

al-Farghānī, Abu'l-'Abbās Aḥmad ibn Muḥammad. *Elementa Astronomica (Jawāmi' 'ilm al-nujūm)*, ed. J. Golius. Amsterdam: Jansonius à Waasberge, 1669.

Friedmann, Yohanan. 'al-Manṣūra', in *EI2*, vol. 6, pp. 439–40.

Gautier Dalché, Patrick. '*Epistola Fratrum Sincerorum in Cosmographia*: Une traduction latine inédite de la quatrième *Risāla* des Ikhwān al-Ṣafā'', *Revue d'histoire des textes*, vol. 18 (1988), pp. 137–67.

Golden, P. B. 'Toghuzghuz', in *EI2*, vol. 10, pp. 555–57.

—— 'Alans', in *EI3*.

Gutas, Dimitri. *Greek Thought, Arabic Culture*. London: Routledge, 1998.

Hamdani, Abbas. 'The Arrangement of the *Rasā'il Ikhwān al-Ṣafā'*', in *The Ikhwān al-Ṣafā' and their* Rasā'il. *An Introduction*, ed. Nader El-Bizri. Oxford: Oxford University Press–The Institute of Ismaili Studies, 2008, pp. 83–100.

Hartmann, R., and D. M. Dunlop. 'Baḥr al-Hind', in *EI2*, vol. 1, pp. 930–31.

Heck, Paul L. *The Construction of Knowledge in Islamic Civilization. Qudāma b. Ja'far and his* Kitāb al-Kharāj wa-ṣinā'at al-kitāba. Leiden: Brill, 2002.

Honigmann, Ernst. *Die sieben Klimata*. Heidelberg: Carl Winter, 1929.

Ibn al-'Adīm, Kamāl al-Dīn 'Umar ibn Aḥmad. *Bughyat al-ṭalab fī ta'rīkh Ḥalab*, ed. S. Zakkār. Beirut: Dār al-Fikr, 1988.

Ibn Khurradādhbih, 'Ubayd Allāh ibn 'Abd Allāh. *Kitāb al-Masālik wa'l-mamālik*, ed. M. J. de Goeje. *Bibliotheca Geographorum Arabicorum*, vi. Leiden: Brill, 1889.

Ibn Qutayba, 'Abd Allāh ibn Muslim. *Faḍl al-'arab wa'l-tanbīh 'alā 'ulūmhā*, ed. Maḥmūd Khāliṣ. Abu Dhabi: al-Majma' al-Thaqāfī, 1998.

Ibn Shaddād, Bahā' al-Dīn Yūsuf ibn Rāfī. *Al-A'lāq al-khaṭīra fī dhikr umarā' al-Shām wa'l-Jazīra*, ed. Y. Z. 'Abbāra. Damascus: Manshūrāt Wizārat al-Thaqāfa, 1991.

Kramers, J. H. 'L'Influence de la tradition iranienne dans la géographie arabe', in *Analecta Orientalia. Posthumous Writings and Selected Minor Works*. Leiden: Brill, 1954, vol. 1, pp. 147–56.

al-Khwārizmī, Muḥammad ibn Mūsā. *Kitāb Ṣūrat al-arḍ*, ed. H. von Mžik as *Das Kitāb ṣūrat al-arḍ des Abū Ǧa'far Muḥammad Ibn Mūsā al-Ḫuwārizmī*. Leipzig: Harrassowitz, 1926.

Kunitzsch, Paul. *Untersuchungen zur Sternnomenklatur der Araber*. Wiesbaden: Otto Harrassowitz, 1961.

Le Strange, G. *The Lands of the Eastern Caliphate*. Cambridge: Cambridge University Press, 1930.

Massé, R. 'Farīdūn', in *EI2*, vol. 2, p. 798.

al-Mas'ūdī, 'Alī ibn al-Ḥusayn. *Kitāb al-Tanbīh wa'l-ishrāf*, ed. M. J. de Goeje. *Bibliotheca Geographorum Arabicorum*, viii. Leiden: Brill, 1894; repr. 1967.

McCrindle, J. W. *Ancient India as Described by Ptolemy*. London: Trübner, 1885.

Miquel, André. *La géographie humaine du monde musulman jusqu'au milieu du 11e siècle*. 4 vols. Paris: Éditions de l'EHESS, 1973–1988.

Montgomery, James. 'Serendipity, Resistance, and Multivalency', in *On Fiction and* Adab *in Medieval Arabic Literature*, ed. P. F. Kennedy. Wiesbaden: Harrassowitz, 2005, pp. 177–230.

al-Muqaddasī, Muḥammad ibn Aḥmad. *Kitāb Aḥsan al-taqāsīm fī ma'rifat al-aqālīm*, in *Bibliotheca Geographorum Arabicorum*, ed. M. J. de Goeje. 8 vols. Leiden: Brill, 1967, vol. 3.

Nallino, C. A. 'Al-Ḫuwārizmī e il suo rifacimento della Geografia di Tolomeo', in *Atti della Real Accademia dei Lincei*, anno 291, serie V, *Memorie delle Classe di Scienze Morali* (1894), vol. 2, pp. 4–53.

Nasr, Seyyed Hossein. *An Introduction to Islamic Cosmological Doctrines*. New York: State University of New York Press, 1993.

—— *The Islamic Intellectual Tradition in Persia*. Richmond: Curzon Press, 1996.

Pedersen, Olaf. *A Survey of the Almagest*. Odense University Press, 1974, rpr., New York: Springer Science, 2010.

Poonawala, Ismail K. 'Why We Need an Arabic Critical Edition with an Annotated English Translation of the *Rasā'il Ikhwān al-Ṣafā*", in *The Ikhwān al-Ṣafā and their* Rasā'il. *An Introduction*, ed. N. El-Bizri. Oxford: Oxford University Press–The Institute of Ismaili Studies, 2008, pp. 33–57.

Ptolemy. *The Geography*, tr. E. L. Stevenson. New York Public Library 1932, rpr., London: Constable, 1991.

Rippin, A., '*Āya*', in *EI3*.

Savant, Sarah Bowen. 'Isaac as the Persian Ishmael: Pride and the Pre-Islamic Past in Ninth and Tenth Century Islam', *Comparative Islamic Studies*, vol. 2, 1 (2006), pp. 5–25.

Sezgin, Fuat. *Geschichte des arabischen Schrifttums. Mathematische Geographie und Kartographie im Islam*, vol. 10. Frankfurt: Institut für Geschichte der Arabisch-Islamischen Wissenschaften, 2000.

—— *Mathematical Geography and Cartography in Islam and Their Continuation in the Occident*, vol. 2. Frankfurt: Institute for the History of Arabic-Islamic Science, 2007.

Silverstein, Adam. 'The Medieval Islamic Worldview: Arabic Geography in its Historical Context', in *Geography and Ethnography: Perceptions of the World in Pre-Modern Societies*, ed. K. A. Raaflaub and R . J. A. Talbert. Chichester: Wiley–Blackwell, 2010, pp. 273–90.

Sizgorich, Thomas. 'Narrative and Community in Islamic Late Antiquity', *Past & Present*, vol. 185, 1 (2004), pp. 9–42.

Tibbetts, G. R. 'The Beginnings of a Cartographical Tradition', in *History of Cartography*, ed. J. B. Harley and D. Woodward. 6 vols. Chicago: Chicago University Press, 1992, vol. 2, pp. 90–107.

Tibbetts, G. R., and Sh. Toorawa. 'Zābadj', in *EI2*, vol. 11, pp. 367–69.

Travaglia, P. *Causality and Intentionality: the Doctrine of the Rays in al-Kindi*. Micrologus Library 3. Florence: Galluzzo, 1999.

Van Donzel, E. J., and A. B. Schmidt. *Gog and Magog in Early Christian and Islamic Sources*. Leiden: Brill, 2009.

Wright, Owen (ed.). *On Music. An Arabic Edition and English*

Translation of Epistle 5. Oxford: Oxford University Press–The Institute of Ismaili Studies, 2010.

al-Yaʿqūbī, Aḥmad ibn Abī Yaʿqūb. *Kitāb al-Buldān*, ed. W. T. Juynboll. *Bibliotheca Geographorum Arabicorum,* vii. Leiden: Brill, 1861.

Zadeh, Travis. *Mapping Frontiers Across Medieval Islam. Geography, Translation and the ʿAbbasid Empire*. London: I.B. Tauris, 2011.

Index

The index does not include the appendices.

رَسَائِلُ إِخْوَانِ الصَّفَاءِ وخُلَّانِ الوَفَاءِ

رَسَائِلُ إِخْوَانِ الصَّفَاءِ وخُلاَّنِ الوَفَاءِ

(٤)

الرِسالَةُ الرابِعَةُ

في الجغرافيا

مِن القِسمِ الأَوَّلِ في العُلومِ الرياضيَّةِ التَعْليميَّةِ

حَقَّقَها وتَرجَمَها

إغناثيو سانشِزْ بالتعاون مع جايمس مُنتغُمري

دَار نَشْر جامِعة أُكسْفُورْدْ

بالتَعاونِ مع مَعْهَد الدِّراساتِ الإسْماعيلّيَة في لَنْدَن

٢٠١٤

رسالة جُغْرافِيَا يعني صورة الأرض والأقاليم وهي الرابعة من القسم الأوّل من الرياضيّات

بسم الله الرحمن الرحيم[١]

من أجل أنّ مذهب[٢] إخواننا أيّدهم الله وإيّانا بروح منه هو النظر[٣] في جميع الموجودات والبحث عن مبادئها[٤] وعن علّة وجدانها[٥] وعن مراتب[٦] نظامها والكشف عن كيفيّة ارتباط[٧] معلولاتها بعلّاتها[٨] بإذن بارئها[٩] جلّ ثناؤه احتجنا أن نذكر[١٠] حال الأرض وكيفيّة صورتها وسبب وقوفها (١٥٩) في مركز العالم. وذلك أنّ المعرفة بحالها وكيفيّة[١١] وقوفها في الهواء من العلوم الشريفة لأنّ عليها وقوف أجسامنا ومنها بدء كون أجسادنا

---

(١) سقط من أ، ر، غ: بسم الله الرحمن الرحيم

(٢) مذهب: غرض [ن]

(٣) منه هو النظر: منه النظر [ع]، مذهبهم النظر [ف]

(٤) سقط من ع: مبادئها

(٥) وجدانها: وجداتها [أ، ط]. سقط من ك: وجدانها

(٦) سقط من ن: والبحث عن مبادئها وعلّة وجدانها عن مراتب

(٧) عن كيفيّة ارتباط: عن ارتباط [ع]

(٨) بعلّاتها: بعللها [ر، ز، ع، غ، ك، ن، ل]

(٩) زيادة في ر: الذي هو معلّ علّة العلل ومبدع المبدعات لا من شيء

(١٠) زيادة في ع: في هذه الرسالة

(١١) وكيفيّة: وبمعرفة كيفيّة [خ]، وبكيفيّة [ز]

ونشوؤها ومادّة بقائها وإليها عودها عند مفارقتها نفوسنا[١] وأيضاً فإنّ النظر في هذا العلم[٢] يكون سبباً لترقّي همم نفوسنا إلى عالم الأفلاك[٣] مسكن العلّيّين وبكثرة (ب ٣٥) جولان أفكارنا في محالّ الروحانيّين[٤] وكثرة ذكرنا للعالم[٥] أعني عالم الأفلاك[٦] يكون سبباً لانتباه نفوسنا من نوم الغفلة ورقدة الجهالة ويدعوها ذلك إلى الانبعاث من عالم الكون والفساد إلى عالم البقاء والدوام ويرغّبها[٧] في الرحلة من عالم الأجسام وجوار الشياطين[٨] إلى عالم الأرواح وجوار الملائكة المقرّبين.

وقد ذكرنا في هذه الرسالة[٩] طرفاً من كيفيّة صورة الأرض وصفة الربع المسكون منها وما فيه من الأقاليم[١٠] ومن البحار والجبال والبراريّ والأنهار[١١] والمدن ليكون طريقاً للمبتدئين

---

(١) نفوسنا: نفوسها [ف]

(٢) العلم: العالم مع التصحيح «العلم» [ط]، العالم [ك]

(٣) عالم الأفلاك: عالم الأرواح [ع]. زيادة في ر: التسعة وهو فلك البروج والمحيط بها وهي

(٤) في محالّ الروحانيّين: في عالم الأفلاك [ع]

(٥) للعالم: في العالم [ر، ط]

(٦) للعالم أعني عالم الأفلاك: في عالم الأفلاك مسكن العلّيّين [خ]، في عالم الأفلاك [ز، غ، ك، ل]، لعالم الأفلاك [ف]. سقط من ز: أعني عالم الأفلاك. سقط من ع، ن: وكثرة ذكرنا للعالم أعني عالم الأفلاك. الأفلاك: الأرواح مع التصحيح «الأفلاك» [ط]

(٧) ويرغّبها: ونرغّبها [ج]، وترفعها [خ]، ورغبتنا [ز]، وترغّبها [ف]

(٨) جوار الشياطين: جوار عالم الشياطين [ع]. الشياطين: الشيطان [خ، غ]

(٩) هذه الرسالة: هذا الكتاب [غ]

(١٠) زيادة في ز، غ، ك: السبعة

(١١) سقط من خ: والأنهار

بالنظر في علم الهيئة وتركيب الأفلاك وطوالع البروج ودوران الكواكب وليقرب[١] تصوّرها في أفكار المتعلّمين ويسهل تأمّلها للمتفكّرين ﴿فِي مَلَكُوتِ ٱلسَّمٰوَاتِ وَٱلْأَرْضِ﴾ الذين يقولون ﴿رَبَّنَا مَا خَلَقْتَ هٰذا بَاطِلاً سُبْحَانَكَ فَقِنَا عَذَابَ ٱلنَّارِ﴾[٢] (١٦٠)

ذكر الأقاليم وما في الربع المسكون من الأرض والجبال والبحار والبراريّ والأنهار والمدن وما في البحار من الجزائر والمدن

وقبل وصفها نحتاج أن نذكر صفة الأرض وجهاتها الستّة وكيف[٣] وقوفها في الهواء. الجهات هي الشرق والغرب[٤] والجنوب والشمال والفوق والأسفل.[٥] والشرق من حيث تطلع الشمس والغرب من حيث تغيب[٦] الشمس والجنوب من حيث مدار سُهَيْل والشمال من حيث مدار الجَدْي والفَرْقَدَيْن[٧] والفوق

(١) وليقرب: ليقرب [أ]، في تقريب [ج]، ويقرب [خ، ر، ز، ط، غ، ل، ن]

(٢) زيادة في ز، ع، ك: وقال تعالى ﴿وَفِي ٱلْأَرْضِ آيَاتٌ لِلْمُوقِنِينَ﴾ وقال تعالى ذكره ﴿وَكَذٰلِكَ نُرِي إِبْرَاهِيمَ مَلَكُوتَ ٱلسَّمٰوَاتِ وَٱلْأَرْضِ وَلِيَكُونَ مِنَ ٱلْمُوقِنِينَ﴾. زيادة في هامش ن: ﴿وَفِي ٱلْأَرْضِ آيَاتٌ لِلْمُوقِنِينَ ۝ وَفِي أَنْفُسِكُمْ أَفَلَا تُبْصِرُونَ﴾ ﴿وَكَذٰلِكَ نُرِي إِبْرَاهِيمَ مَلَكُوتَ ٱلسَّمٰوَاتِ وَٱلْأَرْضِ وَلِيَكُونَ مِنَ ٱلْمُوقِنِينَ﴾ سبحان ربّك ربّ العزّة عمّا يصفون وسلام على المرسلين والحمد لله ربّ العالمين

(٣) وكيف: وكيفيّة [ر، ز]

(٤) الشرق والغرب: المشرق والمغرب [أ، غ، ل]

(٥) والفوق والأسفل: وأسفل وفوق [ع]، والفوق والتحت [غ]

(٦) تغيب: تغرب [ط، غ، ك، ل]

(٧) سقط من ف، ع: والفرقدين

هو ممّا يلي السماء والأسفل[1] ممّا يلي مركز الأرض.[2]

الأرض جسم مدوّر[3] مثل الكرّة وهي واقفة في الهواء بإذن الله تعالى بجميع جبالها وبحارها وبراريّها وعمرانها وخرابها والهواء محيط بها من جميع جهاتها شرقها وغربها وجنوبها وشمالها ومن ذا الجانب ومن ذاك الجانب وبعد الأرض من السماء من جميع جهاتها[4] متساو وأعظم دائرة على بسيط الأرض ٢٠٤٠٠ ميلاً[5] يكون[6] ٦٨٠٠ فرسخاً[7] وقطر هذه الدائرة هو قطر الأرض[8] ٦٥٠٠ ميلاً[9] يكون ٢١٧٠[10] فرسخاً

---

(١) الأسفل: أسفل [ع]، التحت [ف]

(٢) مركز الأرض: كرّة الأرض [ز]، المركز [ن]، ركن الأرض [غ]. سقط من ج: والشرق من حيث تطلع الشمس والغرب من حيث تغيب الشمس والجنوب من حيث مدار سهيل والشمال من حيث مدار الجدي والفرقدين؛ والفوق هو ممّا يلي السماء والأسفل ممّا يلي مركز الأرض

(٣) مدوّر: مدوّر الشكل [أ]

(٤) من جميع جهاتها: وسائر نواحيها [ف]. سقط من غ: شرقها وغربها وجنوبها وشمالها ومن ذا الجانب ومن ذاك الجانب وبعد الأرض من السماء من جميع جهاتها

(٥) ٢٠٤٠٠ ميلا: ٣٠٤٠٠ ميلا [ج، خ]، ٢٤٠٠٠ ميلا [ع]، عشرين ألف وأربعمائة ميلا [ف]، ٢٠٤٣١ ميلا [ن].

(٦) يكون: فيكون [ن]. سقط من ر، ز، ط، غ، ك، ل: يكون

(٧) ٦٨٠٠ فرسخا: فرسخ [خ]، ٢٩٠٠ فرسخا [غ، ف]، ٦٧١٠ فرسخ [ن]. سقط من ع: ٦٨٠٠ فرسخا. زيادة في ع: كلّ ثلثة أميال فرسخ.

(٨) سقط من ط: هو قطر الأرض

(٩) ٦٥٠٠ ميلا: ٦٥٠١ ميلا [ز، ط، غ، ك، ل]، ستّة آلاف وخمسمائة ميلا ٦٥٠٠ [ف]، ٦٤٠١ ميلا [ن]

(١٠) ٢١٧٠: ٢١٩٧ [ن]، ٢١٦٧ [ج، خ، ز، ط، غ، ك، ل]، ألفي ومائة وسبعة وستّين [ف]

بالتقريب[١] ومركزها هي نقطة متوهّمة[٢] في عمقها على نصف القطر وبعدها من ظاهر سطح الأرض والبحر[٣] من جميع الجهات متساو[٤] وليس شيء من ظاهر سطح الأرض من جميع جهاتها هو أسفل الأرض كما يتوهّم كثير من الناس ممّن ليس له رياضة في علم الهندسة.[٥] وذلك أنّهم يتوهّمون ويظنّون بأنّ سطح الأرض من الجانب المقابل لموضعنا[٦] هو أسفل الأرض[٧] وأنّ الهواء المحيط[٨] بذلك الجانب هو أيضاً أسفل من الأرض وأنّ النصف من فلك القمر المحيط بالهواء هو أيضاً أسفل من الهواء وهكذا سائر طبقات (١٦١) الأفلاك كلّ واحد أسفل من الآخر حتّى يلزم من هذا الوهم أنّ أسفل السافلين هو

---

(١) ٦٥٠٠ ميلا يكون ٢١٧٠ فرسخا بالتقريب: ٧٦٣٦ ميلا وأربعة أجزاء من أحد عشر جزءا من ميل وهو ٢٩٤٩ فرسخا وخمسة عشر جزءا من ثلثة وثلاثين جزءا من فرسخ [ع]، ٦٤٥١ ميلا و٣١٦٧ فرسخا [ر]

(٢) نقطة متوهّمة: نقطة وهميّة [ع]

(٣) والبحر: وسطح البحار [ر، ع]، ومن سطح البحار [ز]، ومن سطح البحر [ط، غ، ك، ل]

(٤) زيادة في ز، ع، ك: لأنّ الأرض بجميع البحار التي على ظهرها كرّة واحدة

(٥) رياضة في علم الهندسة: نظر بالرياضة في الهندسة وعلم الهيئة [ع]، رياضة بالنظر في علم الهندسة والهيئة [ز، ط]، رياضة بالنظر [ك]. زيادة في ل: والهيئة

(٦) لموضعنا: لموضعهم [ع]، لموضعها [غ]

(٧) سقط من خ: كما يتوهّم كثير من الناس ممّن ليس له رياضة في علم الهندسة وذلك أنّهم يتوهّمون ويظنّون بأنّ سطح الأرض من الجانب المقابل لموضعنا هو أسفل الأرض. سقط من ر: الأرض

(٨) سقط من ك: وذلك أنّهم يتوهّمون ويظنّون بأنّ سطح الأرض من الجانب المقابل لموضعنا هو أسفل الأرض وأنّ الهواء المحيط

نصف الفلك المحيط الذي هو أعلى العلّيّين في دائم الأوقات.[١]

وليس الأمر كما توهّموا لأنّ هذا رأي ضعيف[٢] يعتقده الإنسان من الصبى[٣] من غير رويّة[٤] ولا برهان فإذا ارتاض الإنسان في علم الهيئة[٥] تبيّن له أنّ الأمر بخلاف ما توهّم قبل ذلك. وذلك أنّ أسفل الأرض بالحقيقة[٦] هو عمق باطنها ممّا يلي مركزها من أيّ جانب كان من الأرض لأنّ مركز الأرض هو أسفل السافلين[٧] وأما سطحها الظاهر المماسّ[٨] للهواء[٩] من جميع الجهات هو الفوق والهواء المحيط هو فوق الأرض أيضاً من جميع الجهات وفلك القمر فوق[١٠] الهواء وفلك عطارد فوق فلك القمر وعلى هذا القياس سائر الأفلاك واحد فوق الآخر إلى

(١) سقط من ج، خ، ز، ط، ن، ع، غ، ف: حتّى يلزم من هذا الوهم أنّ أسفل السافلين هو نصف الفلك المحيط الذي هو أعلى علّيّين في دائم الأوقات

(٢) سقط من ج، خ، ر، ز، ع، غ: ضعيف

(٣) زيادة في ز، ط، ع، ك، ل: بالتوهّم. سقط من ك: من الصبى

(٤) زيادة في غ: ولا فكرة

(٥) الهيئة: الهندسة [ر،ع، ن]، الهيئة والهندسة [ز، ط، ك]. زيادة في هامش ع: ونظر في علم الهيئة

(٦) زيادة في ز، ك: هو نقطة وهميّة في عمق الأرض على نصف قطرها وهو الذي يسمّى مركز العالم

(٧) السافلين : سافلين [ف، ن]. وذلك انّ أسفل الأرض بالحقيقة هو عمق باطنها ممّا يلي مركزها من أيّ جانب كان من الأرض لأنّ مركز الأرض هو أسفل السافلين: واعلم أيّدك الله وإيّانا بروح منه بأنّ أسفل السافلين بالحقيقة هو نقطة وهميّة في عمق الأرض على نصف قطرها وهو الذي يسمّى مركز العالم [ع]

(٨) سطحها الظاهر المماسّ: السطح الظاهر المباشر [أ]

(٩) زيادة في ز، ط، ع، ك: وسطح البحار

(١٠) زيادة في أ، ر، غ: فلك

الفلك التاسع الذي هو فوق كلّ فوق وهو أعلى العلّيّين ومقابله مركز الأرض أسفل السافلين.

واعلم يا أخي أنّ الإنسان أيّ موضع وقف على سطح الأرض من شرقها أو غربها أو جنوبها أو شمالها أو من ذاك الجانب أو من ذا الجانب فموقفه(١) أبداً يكون(٢) فوق الأرض ورأسه إلى فوق ممّا يلي السماء ورجلاه أسفل ممّا يلي مركز(٣) الأرض وهو يرى من السماء نصفها والنصف الآخر تستر عنه حدبة الأرض فإذا انتقل الإنسان من موضع إلى موضع(٤) ظهر له من السماء مقدار ما خفي عنه من الجهة الأخرى فذلك المقدار في كلّ تسعة عشر فرسخاً درجة واحدة.(٥) (١٦٢)

ذكر سبب وقوف الأرض في وسط الهواء

وأمّا سبب وقوف الأرض في وسط الهواء(٦) ففيه أربعة أقاويل منها ما قيل إنّ سبب وقوفها هو جذب الفلك لها من جميع الجهات بالسويّة فوجب لها الوقوف في الوسط لمّا

---

(١) فموقفه: فهو فوقه [ر]، فوقه [خ، ل، ن]، فقدمه [ط]، فوقوفه [غ، ف]

(٢) فموقفه أبدا يكون: وقوفه حيث كان فقدمه [ز]، فإنّه حيث كان يكون [ع]، وقوفه حيث كان فقدمه أبدا يكون [ك]

(٣) سقط من خ، غ: مركز

(٤) من موضع إلى موضع: من ذلك الموضع إلى الموضع الآخر [ج، خ، ر، ز، ط، غ، ف، ك، ل، ن]

(٥) سقط من ج، خ، ر، ط، ع، غ، ك، ل، ن: واحدة. زيادة في ز، ط، ع، ك، ل، ن: كلّ فرسخ ثلاثة أميال وكلّ ميل أربعة آلاف ذراع وكلّ ذراع ستّ قبضات وكلّ قبضة أربع أصابع وكلّ إصبع ستّ شعيرات

(٦) الهواء: السماء [ن]

تساوى[١] قوّة الجذب[٢] من جميع الجهات ومنها ما قيل إنّه الدفع مثل ذلك[٣] فوجب لها الوقوف في الوسط[٤] لمّا تساوى قوّة الدفع من جميع الجهات[٥] ومنها ما قيل إنّ سبب[٦] وقوفها في الوسط هو جذب المركز بجميع أجزائها[٧] من جميع الجهات[٨] إلى الوسط لأنّه لمّا كان مركز الأرض مركز الفلك أيضاً وهو[٩] مِغْنَاطِيس الأثقال يعني مركز الأرض في أجزاء[١٠] الأرض[١١] لمّا كانت كلّها ثقيلة فانجذب إلى المركز وسبق جزء واحد وحصل في المركز وصار[١٢] ووقف باقي الأجزاء حولها يعني حول النقطة[١٣] وطلب كلّ جزء منها المركز فصارت الأرض بجميع أجزائها كرّة واحدة بذلك السبب ولمّا كان الماء أجزاؤه أخفّ من أجزاء الأرض وقف الماء فوق الأرض ولمّا كانت أجزاء الهواء أخفّ من أجزاء الماء صار فوق الماء والنار ولمّا كانت أجزاؤها أخفّ

(١) لمّا تساوى: لتساوي [ع]، تساوت [غ]
(٢) لمّا تساوى قوّة الجذب: لمّا قوّة الجذب تساوت [ن]
(٣) الدفع مثل ذلك: دفع الفلك لها من جميع الجهات [ع]، مثل [غ]
(٤) سقط من ع: في الوسط
(٥) سقط من خ: ومنها ما قيل إنّه الدفع مثل ذلك فوجب لها الوقوف في الوسط لمّا تساوى قوّة الدفع من جميع الجهات
(٦) سقط من ل: سبب
(٧) أجزائها: جهاتها [خ]. سقط من ل: بجميع أجزائها
(٨) سقط من ع: من جميع الجهات
(٩) سقط من ع: لمّا كان مركز الأرض مركز الفلك أيضا وهو
(١٠) في أجزاء: وأجزاء [ط، غ، ف، ك]
(١١) سقط من خ: في أجزاء الأرض
(١٢) سقط من ج، خ، ر، ز، ع، غ، ك، ل، ن: وصار
(١٣) سقط من ع: يعني حول النقطة

من أجزاء الهواء صارت في العلوّ ممّا يلي فلك القمر. والوجه الرابع ممّا قيل في سبب وقوف الأرض في وسط الهواء هو خصوصيّة الموضع الذي أليق بها[1] (أ ٣٦) وذلك أنّ الباريء[2] تبارك وتعالى جعل لكلّ جسم من الأجسام الكلّيّات يعني النار والهواء والماء والأرض موضعاً مخصوصاً هو أليق الموضع[3] به[4] وهكذا القمر وعطارد والزهرة والشمس والمرّيخ والمشتري وزحل[5] جعل[6] لكلّ واحد منها[7] موضعاً مخصوصاً[8] في فلكه[9] هو ثابت فيه والفلك يديره معه[10] وهذا القول أشبه الأقاويل بالحقّ لأنّ هذه العلّة مستمرّة في ترتيب الأفلاك التسعة والكواكب الثابتة والسيّارة والأركان الأربعة أعني النار والهواء والماء والأرض[11] وذلك أنّ الله تبارك وتعالى جعل[12] لكلّ

---

(١) الموضع الذي أليق بها: الموضع اللائق بها [أ، ر، ز، ط، غ، ف، ك، ل، ن]، الموضع اللائق [ج]، الموضع الأليق بها [خ]

(٢) الباريء: الله [ف]

(٣) الموضع: المواضع [ج، ز، ط، غ، ف، ك، ل، ن]

(٤) أليق الموضع به: أليق به [ع]

(٥) سقط من ع: وهكذا القمر وعطارد والزهرة والشمس والمرّيخ والمشتري وزحل

(٦) سقط من أ، ج، خ، ر، ع، غ، ن: جعل.

(٧) واحد منها: كوكب في الفلك [ع]، واحد منهم [ف]

(٨) موضعا مخصوصا: موضوع مخصوص [ر، ن، ل]، موضع [غ]، موضوع مخصوص به [ف]

(٩) سقط من ع: في فلكه

(١٠) سقط من ع: والفلك يديره معه

(١١) في ترتيب الأفلاك التسعة والكواكب الثابتة والسيّارة والأركان الأربعة أعني النار والهواء والماء والأرض: في جميع الأجسام وجميع الموجودات [ع]

(١٢) زيادة في ع: بواجب حكمته

موجود من الموجودات[١] موضعاً يختصّ[٢] به دون سائر المواضع[٣] (١٦٣) ورتبة معلومة[٤] هي أليق به[٥] من سائر المراتب.[٦]

ذكر الأرض وقسمة أرباعها[٧]

وسط الأرض نصفها مغطّى[٨] بماء البحر الأعظم المحيط[٩] والنصف الآخر مكشوف[١٠] مثلها مثل بيضة غائصة نصفها في الماء والنصف الآخر نات[١١] من الماء[١٢] ومن هذا النصف المكشوف نصف منه خراب ممّا يلي الجنوب من خطّ الاستواء والنصف الآخر الذي[١٣] هو الربع المسكون ممّا يلي الشمال من

(١) زيادة في ز: أو. سقط من ع، ف: من الموجودات
(٢) يختصّ: مختصّ [ط]، مخصوصا [ف]
(٣) سقط من ف: به دون سائر المواضع
(٤) موضعا يختصّ به دون سائر المواضع ورتبة معلومة: موضعا يختصّ به دون سائر المواضع المرتبة معلومة [أ]، موضعا مخصوصا ورتبة معلومة [ع، ن]
(٥) سقط من ج، خ: دون سائر المواضع ورتبة معلومة هي أليق به
(٦) سقط من ع: من سائر المراتب. زيادة في ع: كما ذكر جلّ ثناؤه ﴿وَمَا مِنَّا إِلَّا لَهُ مَقَامٌ مَعْلُومٌ ۝ وَإِنَّا لَنَحْنُ الصَّافُّونَ ۝ وَإِنَّا لَنَحْنُ الْمُسَبِّحُونَ﴾ يعني الملائكة
(٧) زيادة في ع: واعلم أيّها الأخ البارّ الرحيم أيّدك الله وإيّانا بروح منه أنّ. زيادة في ن: فنقول اعلم أيّها الأخ أنّ سطح
(٨) مغطّى: مغطاة [ج، خ، ز، ط، ف، ك، ل، ن]
(٩) مغطّى بماء البحر الأعظم المحيط: يحيط به البحر الأعظم [ع]
(١٠) زيادة في ف: بائن
(١١) نات: ناتي [أ، ج، ط]، باين [ز، غ، ن]، نار [ك]
(١٢) سقط من خ: والنصف الآخر نات من الماء
(١٣) سقط من ع، غ: الذي

خطّ الاستواء[١] وخطّ الاستواء هو خطّ متوهّم ابتداؤه من المشرق إلى المغرب تحت مدار رأس برج الحمل والليل والنهار على ذلك الخطّ أبداً متساويان والقطبان هنالك ملازمان للأفقين أحدهما ممّا يلي مدار سهيل في الجنوب والآخر في الشمال ممّا يلي مدار الجدي[٢] وهذا مثال ذلك.[٣]

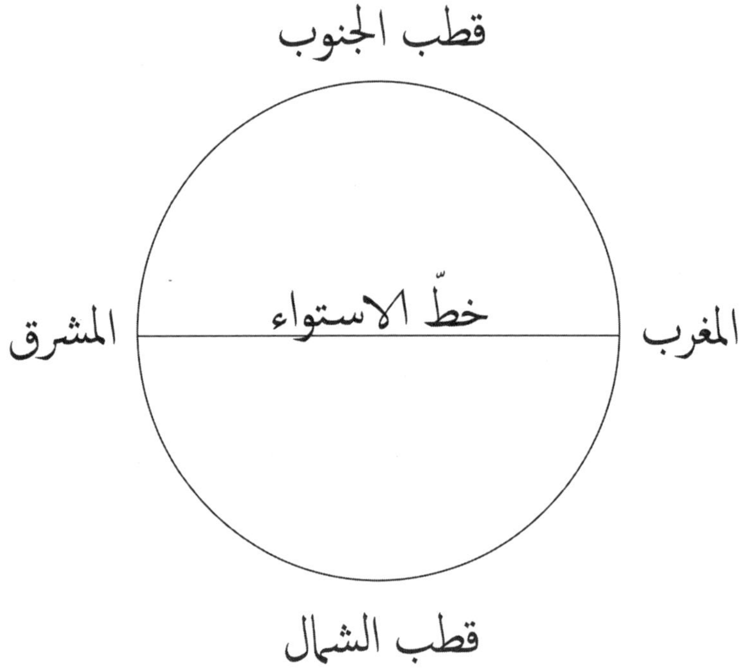

صفة الربع المسكون من الأرض

وفي هذا الربع الشماليّ المسكون سبعة أبحار كبار وفي كلّ

(١) سقط من خ: والنصف الآخر الذي هو الربع المسكون ممّا يلي الشمال من خطّ الاستواء

(٢) زيادة في ع، غ: والفرقدين. زيادة في ف: والقوس والفرقدين

(٣) وهذا مثال ذلك: وهذا أمثال في ابتداء الصفحة الآتية [ز]، وقد مثلنا له لك مثالا ليكون أقرب لفهم الناظرين وأثبت في نفوس المبتدئين وهو هذا [غ]

بحر منها عدّة جزائر تكسير كلّ جزيرة منها عشرون فرسخاً إلى مائة فرسخ إلى ألف فرسخ فمنها بحر الروم وفيه نحو من خمسين جزيرة ومنها (١٦٤) بحر الصقالبة وفيه نحو من ثلاثين[١] جزيرة ومنها بحر جرجان وفيه خمس جزائر[٢] ومنها بحر القلزم وفيه خمس عشرة جزيرة ومنها بحر فارس وفيه سبع جزائر[٣] ومنها بحر السند والهند وفيه نحو من ألف جزيرة ومنها بحر الصين وفيه نحو من مائتي جزيرة.

وفي هذا الربع أيضاً نحو من خمس عشرة بحيرة صغار تكسير كلّ واحد منها من عشرين فرسخاً إلى مائة فرسخ إلى ألف فرسخ منها مالح ومنها عذب وأمّا بحر المغرب وبحر ياجوج وماجوج وبحر الزنج وبحر الزابَج[٤] والبحر الأخضر والبحر المحيط فخارج عن هذا الربع المسكون وكلّ واحد من هذه الأبحر شعبة[٥] وخليج من البحر المحيط وكلّها مالح.

وفي هذا الربع أيضاً مقدار مائتي جبل طوال منها ما طوله من عشرين فرسخاً إلى مائة فرسخ[٦] إلى ألف فرسخ مختلفة

(١) ثلاثين: خمسين [ف]

(٢) خمس جزائر: جزائر كثيرة [ع]

(٣) سقط من ر: ومنها بحر القلزم وفيه خمس عشرة جزيرة ومنها بحر فارس وفيه سبع جزائر

(٤) الزابج: الرابح [أ]، الراىح [ج، ع، غ]، الرابج [ر، ط]، الزانج [ك]، الزانح [ل]، الرائح [ن]. سقط من ف: بحر الزابج

(٥) سقط من أ، ج،ع: شعبة

(٦) زيادة في ف: أو من الجنوب إلى الشمال ومنها ما يتنكّب ما بين المشرق والجنوب ومنها ما يتنكّب ما

الألوان.[١] ومنها ما يمتدّ طوله[٢] من المشرق إلى المغرب أو من الجنوب إلى الشمال ومنها ما يتنكّب ما بين المشرق والجنوب[٣] ومنها ما يتنكّب[٤] ما بين المشرق والشمال[٥] ومنها ما هو بين العمران والمدن والقرى ومنها ما هو في البراريّ والقفار[٦] ومنها ما هو في الجزائر والبحار.

وفي هذا الربع أيضاً مقدار مائتين وأربعين نهراً[٧] طول كلّ نهر منها[٨] من عشرين فرسخاً إلى مائة فرسخ إلى ألف فرسخ ومنها ما جريانه[٩] من المشرق إلى المغرب ومنها ما جريانه من المغرب إلى المشرق[١٠] ومنها[١١] من الجنوب إلى الشمال ومنها[١٢] من الشمال إلى الجنوب ومنها في نكباء[١٣] هذه

(١) زيادة في ع: راسية في الأرض راسخة أصولها شامخة في الهواء. زيادة في ف: وهي راسخة في الأرض طولها شامخة

(٢) ومنها ما يمتدّ طوله: ارتفاعها ممتدّة [ع]، ممتدّه [ف]

(٣) سقط من ف: ومنها ما يتنكّب ما بين المشرق والجنوب

(٤) سقط من خ: ما يتنكّب

(٥) ومنها ما يتنكّب ما بين المشرق إلى الجنوب ومنها ما يتنكّب ما بين المشرق والشمال: وما يتنكّب في الجهات [ع]. سقط من ن: ومنها ما يتنكّب ما بين المشرق والجنوب ومنها ما يتنكّب ما بين المشرق والشمال

(٦) سقط من ف: ومنها ما هو بين العمران والمدن والقرى ومنها ما هو في البراريّ والقفار

(٧) زيادة في ر، ع، ف: طوالا

(٨) طول كلّ نهر منها: منها ما طوله [ع]

(٩) ما جريانه: ما يجري [ع]

(١٠) سقط من ع: ومنها ما جريانه من المغرب إلى المشرق

(١١) زيادة في ع: ما يجري. زيادة في ف: ما هو جيرانه

(١٢) زيادة في ف: ما جيرانه

(١٣) نكباء: منتكب [غ]

الجهات[1] وكلّ هذه الأنهار تبتدئ[2] من الجبال وتنتهي إلى البحار[3] في جريانها أو إلى البطائح والبحيرات وتسقي في ممرّها المدن والقرى والسوادات[4] وما يفضل من مائها[5] ما[6] ينصبّ إلى البحار[7] ويختلط بماء البحار[8] ثمّ يصير بخاراً ويصعد في الهواء[9] يتراكم منها الغيوم وتسوقها الرياح إلى رؤوس الجبال والبراريّ[10] ويمطر هناك ويسقي البلاد[11] ويجري الأودية والأنهار ويرجع[12] (١٦٥) إلى البحار من الرأس[13] ذلك تقدير العزيز العليم.

وفي هذا الربع[14] سبعة أقاليم تحتوي على[15] نحو من سبع

(١) في نكباء هذه الجهات: متنكّب هذه الجهات [ر]، ما يتنكّب عن هذه الجهات [ع]. زيادة في ف: الأربعة
(٢) تبتدئ: يجرى ويبتدئ جريانها [ع]
(٣) زيادة في ك: ويختلط بماء البحر
(٤) والسوادات: والسواد [أ]. زيادة في ع: والمزارع
(٥) من مائها: من ذلك [ع]، فيها من الماء [ف]
(٦) سقط من ج، ط، غ، ك: ما
(٧) البحار: البحر [ط]
(٨) بماء البحار: بماء البحر [ط، ك، ل]، بالماء المالح [ع]، بمائها [ف]
(٩) ثمّ يصير بخارا ويصعد في الهواء: ويذوب ويلطف ويتصاعد في الهواء [ع]
(١٠) الجبال والبراريّ: الجبال وبطونها ووجوه البراريّ [ف]
(١١) البلاد: الأرض [ز، ط، ك]
(١٢) ويرجع: وتعود [ف]. زيادة في ع: ما يفضل من ذلك
(١٣) زيادة في ز، ك: وذلك دأبها في الشتاء والصيف. زيادة في ع: وذلك دأبها طول الدهر الشتاء والصيف. زيادة في ف: كما بدأت
(١٤) زيادة في غ، ف: المسكون
(١٥) سقط من ع: تحتوي على. زيادة في ع: فيها

عشرة ألف[١] مدينة[٢] كباراً يملكها نحو من ألف ملك كلّ هذا في ربع واحد من بسيط الأرض وأمّا ثلاثة أرباعها الباقية فحكمها غير هذا.[٣]

صفة الأقاليم السبعة

الأقاليم هي سبع أقسام خطّت[٤] في الربع المسكون من الأرض كلّ إقليم منها كأنّه بساط مفروش قد مدّ طوله من المشرق إلى المغرب وعرضه من الجنوب إلى الشمال وهي مختلفة الطول والعرض وأطولها وأعرضها[٥] الإقليم الأوّل وذلك أنّ طوله[٦] من المشرق إلى المغرب نحو من ثلاثة آلاف فرسخ وعرضه من الجنوب إلى الشمال نحو من مائة وخمسين فرسخاً وأقصرها طولاً وعرضاً[٧] الإقليم السابع وذلك أنّ طوله من المشرق إلى المغرب[٨] نحو من ألف وخمسمائة فرسخ وعرضه من الجنوب إلى الشمال[٩] نحو من سبعين فرسخاً وأمّا سائر

---

(١) سبع عشرة ألف: عشر ألف [ج]

(٢) مدينة: فرسج [أ]

(٣) زيادة في ف: الحكم. زيادة في ك: والله أعلم

(٤) أقسام خطّت: الأقاليم خطّ [ط]، خطط قسمت [ع]

(٥) سقط من ن: وأعرضها

(٦) سقط من خ: طوله

(٧) سقط من ف: وعرضا

(٨) سقط من خ، ف، ل، ن: إلى المغرب

(٩) سقط من أ، ج، خ، ز، ك، ل، ن: إلى الشمال

الأقاليم فيما بينهما[١] من الطول والعرض[٢] فهذا مثال ذلك[٣] (ب ٣٦).

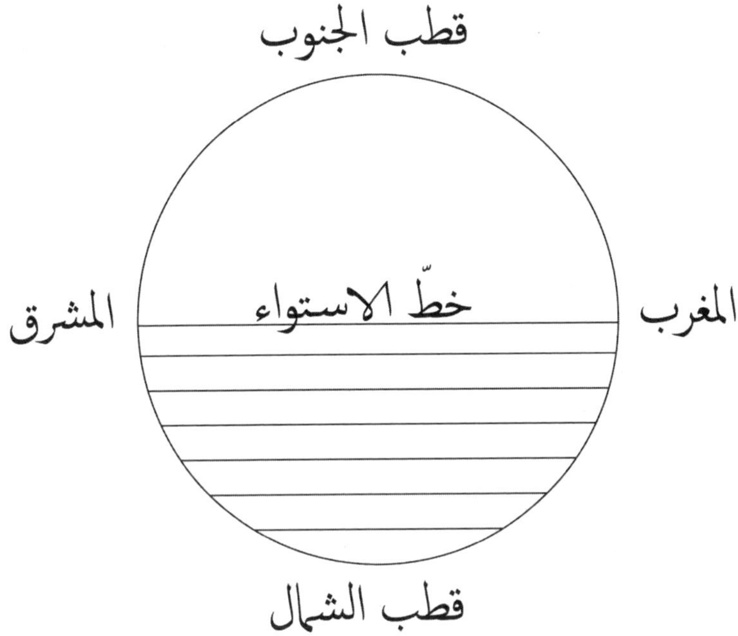

(١٦٦)

اعلم أيّها الأخ البارّ الرحيم أيّدك الله وإيّانا بروح منه أنّ هذه الأقاليم السبعة ليست[٤] أقساما طبيعيّة ولكنّها خطوط وهميّة وضعها الملوك الأوّلون الذين طافوا[٥] الربع المسكون من

---

(١) سقط من أ: بينهما. سائر الأقاليم فيما بينهما: سائر ما بينهما [ع]
(٢) زيادة في ع: فزائد وناقص. زيادة في ف: متسوّى زائد وناقص
(٣) فهذا مثال ذلك: على قياس ذلك وقد تقدّم مثال ذلك [ع]. زيادة في خ: وإنّ إليه المرجع والله أعلم بالصواب وإليه المرجع والمآب. زيادة في ل: والله أعلم الصواب
(٤) زيادة في ز، ط، ك: هي. زيادة في ع: في الموضع المصوّر فيه
(٥) سقط من أ: طافوا

الأرض[1] مثل إسكندر اليونانيّ وتبّع الحميريّ وسليمان بن داود الإسرائيليّ وأفريذون النبطيّ وأرداشير بن بابكان الفارسيّ[2] وغيرهم من الملوك[3] ليعلموا بها[4] حدود البلدان والمسالك والممالك[5] وأمّا ثلاثة أرباعها الباقية فمنعهم من سلوكها الجبال الشامخة والمسالك الوعرة والبحار الزاخرة والأهوية المتغيّرة المفرطة التغيير[6] من الحرّ والبرد والظلمة مثل ما في ناحية الشمال تحت مدار الجدي فإنّ هناك برد مفرط جدّاً لأنّه[7] ستّة أشهر يكون الشتاء هناك ليلاً كلّه[8] فيظلم الهواء ظلمةً

---

(١) زيادة في ز، ك، ل: ليعلم حدود البلدان والممالك والمسالك

(٢) إسكندر اليونانيّ وتبّع الحميريّ وسليمان بن داود الإسرائيليّ وأفريذون النبطيّ وأرداشير بن بابكان الفارسيّ: أفريذون النبطيّ وتبّع الحميريّ وسليمان بن داود الإسرائيليّ والإسكندر اليونانيّ وأرداشير بن بابكان الفارسيّ [ج، ر، ط، غ، ف، ن]؛ أفريذون النبطيّ وتبّع الحميريّ وسليمان بن داود الإسرائيليّ عليهما السلام والإسكندر اليونانيّ وأردشير بابكان الفارسيّ [ز، ك، ل]؛ أفريذون النبطيّ وتبّع الحميريّ وسليمان بن داود الإسرائيليّ والإسكندر اليونانيّ و وأردشير بابكلمن الفارسيّ [خ]

(٣) سقط من خ، ر، ز، ط، غ، ك، ل، ن: وغيرهم من الملوك. سقط من ف: وغيرهم

(٤) ليعلموا بها: ليعلموا أنّها [خ، ط]، فعلموا [ع]

(٥) مثل إسكندر اليونانيّ وتبّع الحميريّ وسليمان بن داود الإسرائيليّ وأفريذون النبطيّ وأرداشير بن بابكان الفارسيّ وغيرهم من الملوك ليعلموا بها حدود البلدان والمسالك والممالك: ليعلم بها حدود البلدان والممالك والمسالك مثل إسكندر اليونانيّ وتبّع الحميريّ وأفريذون القبطيّ وأرداشير الفارسيّ وغيرهم من الملوك فعلموا أنّها حدود البلدان والمسالك والممالك [ع]. البلدان والمسالك والممالك: الممالك والمسالك [ن]

(٦) سقط من ع: المفرطة التغيير

(٧) لأنّه : إلّا أنّ [أ]

(٨) سقط من ع: كلّه

شديدةً[١] ويجمد الماء لشدّة البرد ويتلف النبات والحيوان وفي مقابل هذا الموضع في ناحية الجنوب حيث[٢] مدار[٣] سهيل يكون نهاراً كلّه ستّة أشهر صيفاً[٤] فيحمى الهواء وتصير ناراً سموماً وتحترق الحيوان والنبات من شدّة الحرّ فلا يمكن السكنى[٥] ولا السلوك هناك فأمّا ناحية المغرب فيمنع البحر المحيط السلوك فيها لتلاطم[٦] أمواجه وشدّة ظلماته فأمّا ناحية المشرق فيمنع السلوك هناك الجبال الشامخة فإذا تأمّلت[٧] وجدت الناس محصورين في الربع المسكون من الأرض وليس لهم علم بالثلاثة الأرباع[٨] الباقية.

واعلم أنّ الأرض بجميع ما عليها من الجبال والبحار[٩] بالنسبة إلى سعة الأفلاك ما هي إلّا كالنقطة في الدائرة[١٠] وذلك أنّ في الفلك ألف وتسعة وعشرون كوكباً أصغر كوكب منها مثل الأرض ثماني عشرة مرّةً[١١] وأكثرها[١٢] مائة وسبع مرّات[١٣]

(١) سقط من أ: ظلمة شديدة
(٢) حيث: تحت [ن]
(٣) مدار: يدار [ف]
(٤) سقط من ط، ع، غ: ستّة أشهر صيفا
(٥) السكنى: السكون [خ، ن]
(٦) لتلاطم: المتلاطم [ن]
(٧) زيادة في ع: هنا
(٨) بالثلاثة الأرباع: بثلاثة أرباعها [ع]
(٩) من الجبال والبحار: من الجبال والبراريّ والقفار وسائر البحار [ع]
(١٠) زيادة في غ: صغرى
(١١) سقط من ج: أصغر كوكب منها مثل الأرض ثماني عشرة مرّة
(١٢) زيادة في ج: مثل. زيادة في غ: مثل الأرض
(١٣) مائة وسبع مرّات: مائة وسبعين مرّة [ع]

فلشدّة البعد وسعة الأفلاك نراها كأنّها الدرّ المنثور[١] على بساط أخضر[٢] فإذا فكّر الإنسان في هذه العظمة وتبيّن (١٦٧) له حكمة الصانع وجلاله وعظمته فينتبه[٣] عن نوم الغفلة ورقدة الجهالة ويعلم أنّه ما خلق هذه الأشياء إلّا لأمر عظيم.[٤]

فصل

واعلم يا أخي بأنّ من دخل الدنيا وعاش فيها زماناً طويلاً مشغولاً بالأكل والشرب والنكاح دائباً[٥] في طلب الشهوات والحرص على طلب[٦] جمع المال والأثاث واتّخاذ البنيان وعمارة[٧] العقارات وطلب الرئاسة متمنّياً للخلود فيها تاركاً طلب العلم[٨] غافلاً عن معرفة حقائق الأشياء مهملاً لرياضة النفس متوانياً في استعداد الزاد للرحلة إلى الدار الآخرة[٩] حتّى إذا فني[١٠] العمر وقرب الآجل وجاءت سكرة الموت[١١] التي هي

---

(١) الدرّ المنثور: درّ منثور [ع]، الدرّ المنظوم أو كالمنثور [ف]

(٢) أخضر: أزرق [ر، ع]

(٣) فينتبه: ما ينتبه [غ]. زيادة في ع: نفسه

(٤) زيادة في ج: فتبارك أحسن العالم. زيادة في ز، ع، ك: وذلك قوله تعالى ﴿مَا خَلَقْنَا السَّمٰوَاتِ وَٱلْأَرْضَ . . . إِلَّا بِٱلْحَقِّ﴾. زيادة في ف، ن: وإليه أشار بقوله تعالى ﴿مَا خَلَقْنَا السَّمٰوَاتِ وَٱلْأَرْض. . . إِلَّا بِٱلْحَقِّ﴾

(٥) دائبا: دائما [خ، ز، ف، ل]

(٦) سقط من ج، خ، ط، غ، ف، ك، ل: طلب

(٧) زيادة في ز، ط، ك: الأرض

(٨) العلم: الآخرة [ع]

(٩) للرحلة إلى الدار الآخرة: الموصل إلى الدار الآخرة والرحلة إليها [ع]

(١٠) فني: قضى [غ]

(١١) الموت: الحق والموت [ر]

مفارقة الروح[١] الجسد ثمّ خرج من هذه الدار جاهلاً لم يعرف صورتها ولا فكّر في الآيات التي في آفاقها ولا اعتبر أحوال[٢] موجوداتها ولا تأمّل الأمور المحسوسة التي شاهد فيها.

فمثلهم كمثل قوم دخلوا إلى مدينة ملك عظيم حكيم عادل رحيم قد بناها بحكمته وأعدّ فيها كلّ طرائف من صنعته ما يقصر الوصف عنها إلّا بالمشاهدة لها ووضع فيها مائدة قوتاً للواردين إليها وزاداً للراحلين عنها ثمّ دعا عبيداً له إلى حضرته ليحبوهم بكرامته وأمرهم بالورود إلى تلك المدينة التي[٣] في طريقهم لينظروا إليها[٤] ويبصروا[٥] ما فيها ويتفكّروا في عجائب مصنوعاته ويعتبروا غرائب مصوّراتها ليروَّض[٦] بها نفوسهم فيصيرون برؤيتها[٧] حكماء أخياراً فضلاء فيصلون إلى حضرته ويستحقّون كرامته فوردوها قوم ليلاً فباتوا طول ليلهم مشغولين بالأكل والشرب واللعب واللهو ثمّ خرجوا منها سحراً لا يدرون من أيّ باب دخلوا ولا من أيّها[٨] خرجوا ولا رأوا شيئاً ممّا فيها[٩] من

(١) الروح: النفس [خ، ر، ز، ط، ع، غ، ف، ك، ل، ن]
(٢) سقط من ع: أحوال
(٣) سقط من ج، خ، ز، ط، ف، ك: التي
(٤) إلى تلك المدينة التي في طريقهم لينظروا إليها: إلى المدينة لينظروا إليها في طريقهم [ع]
(٥) ويبصروا: ويتصوّروا [خ، ع]
(٦) ليروّض: ليروّضوا [ر، ع، ف]
(٧) زيادة من ر، ز، ع، غ، ف، ك، ل، ن: ومعرفتها
(٨) من أيّها: من أيّ باب [ع، غ]
(٩) سقط من ف: ممّا فيها

آثار (١٦٨) حكمته وغرائب صنعته ولا انتفعوا[١] بشيء منها[٢] أكثر من تمتّعهم تلك الليلة بالأكل والشرب[٣] حسب فهذا[٤] حكم أبناء الدنيا[٥] الواردين إليها جاهلين الماكثين فيها متحيّرين الراحلين عنها مكرهين المنكرين من أمر الدار الآخرة[٦] كما قال الله جلّ ثناؤه ﴿مَنْ[٧] كَانَ فِي هٰذِهِ أَعْمَى فَهُوَ فِي ٱلْآخِرَةِ أَعْمَى وَأَضَلُّ سَبِيلاً﴾ وقال ذمّاً لهم[٨] ﴿صُمٌّ بُكْمٌ عُمْيٌ فَهُمْ لَا يَعْقِلُونَ﴾ أمر الآخرة.[٩]

فأعيذك أيّها الأخ البارّ الرحيم أن تكون منهم بل كن من الذين مدحهم الله عزّ وجلّ فقال جلّ ثناؤه ﴿تِلْكَ ٱلدَّارُ ٱلْآخِرَةُ نَجْعَلُهَا لِلَّذِينَ لَا يُرِيدُونَ عُلُوّاً فِي ٱلْأَرْضِ وَلَا فَسَاداً وَٱلْعَاقِبَةُ[١٠] لِلْمُتَّقِينَ﴾[١١] وحكي قولهم لمّا تمنّوا[١٢] أبناء[١٣] الدنيا حين

---

(١) انتفعوا: اشتغلوا [ف]

(٢) سقط من ع: منها

(٣) سقط من ر: واللعب واللهو ثمّ خرجوا منها سحرا لا يدرون من أيّ باب دخلوا ولا من أيّها خرجوا ولا رأوا شيئا ممّا فيها من آثار حكمته وغرائب صنعته ولا انتفعوا بشيء منها أكثر من تمتّعهم تلك الليلة بالأكل والشرب. تمتّعهم تلك الليلة بالأكل والشرب: الأكل والشرب وتمتّعهم تلك الليلة [ع]

(٤) حسب فهذا: حسب هممهم الدينيّة فهكذا [ر]

(٥) حكم أبناء الدنيا: مثل أبناء الدنيا [ف]. سقط من أ: الدنيا

(٦) من أمر الدار الآخرة: من أمر الدنيا والآخرة [ف]، لأمن الآخرة [ن]

(٧) من: ومن [ز، غ، ن]

(٨) سقط من ف: ذمّا لهم

(٩) لا يعقلون أمر الآخرة: لا يرجعون يعني بأمر الآخرة [ع، غ]

(١٠) والعاقبة: فالعاقبة [ع]

(١١) سقط من ر: والعاقبة للمتّقين

قالوا(1) ﴿يَا لَيْتَ لَنَا مِثْلَ مَا أُوتِيَ قارُونُ إنّهُ لَذُو حَظٍّ عَظِيم﴾ ﴿وقال ٱلَّذِينَ أُوتُوا ٱلْعِلْمَ﴾ بحقيقة أمر(2) الآخرة(3) ﴿وَيْلَكُمْ ثَوَابُ ٱللهِ خَيْرٌ لِمَنْ آمَنَ وَعَمِلَ صَالِحاً(4) وَلَا يُلَقَّاهَا﴾ ﴿إِلَّا ٱلَّذِينَ صَبَرُوا(5) وَمَا يُلَقَّاهَا إِلَّا ذُو حَظٍّ عَظِيمٍ﴾(6). وفّقك الله أيّها الأخ(7) للسداد والرشاد.(8)

وإذ قد فرغنا من ذكر الأرض ووصفها وربعها المسكون فنريد أن نذكر الأقاليم السبعة(9) ونبيّن حدودها(10) طولاً وعرضاً وما في

---

(١٢) تمنّوا: تمنّى [خ، غ، ن]، تمتّوا [ر، غ]. وحكي قولهم لمّا تمنّوا: وحكي عن قولهم لمّا تمنّى [ع]، وحكي قول أهل الدنيا لمّا تمنّوا [ل]، وحكي قولهم لمّا تمتّوا الموت [ف]

(١٣) أبناء: أعرض [ز، ط، ك، ل]

(١) سقط من ن: حين قالوا

(٢) أمر: حال [غ]

(٣) سقط من ع: بحقيقة أمر الآخرة. سقط من ك: الآخرة. زيادة في خ: والإيمان

(٤) سقط من أ: لمن آمن وعمل صالحاً

(٥) إلّا الذين صبروا: إلّا الصابرون [خ، ر، ط، ع، ف]، إلّا الصابرون الذين صبروا [ن]، إلّا الصابرون وما يلقيها إلّا الذين صبروا [ز]، إلّا الصابرون ولا يلقيها إلّا الذين صبروا [ل]، إلّا الصابرون وقال الله تعالى وما يلقّاها إلّا الذين صبروا [ك]

(٦) سقط من خ، ر: ولا يلقّاها إلّا الذين صبروا وما يلقّاها إلّا ذو حظّ عظيم

(٧) وفّقك الله أيّها الأخ: وفّقك الله وإيّانا للسداد والرشاد [ز]. سقط من غ: وفّقك الله أيّها الأخ. زيادة في خ، ر: البارّ الرحيم.

(٨) للسداد والرشاد: للسداد وهدينا وإيّاك للرشاد وجميع إخواننا حيث كانوا في البلاد [ز]، للسداد وهداك الرشاد وجميع إخواننا حيث كانوا في البلاد [ج، خ، ط، ك]، للسداد وهداك إلى سبيل الرشاد وجميع إخواننا حيث كانوا في البلاد [ل]. والرشاد: وهداك للرشاد وجميع إخواننا حيث كانوا في البلاد [ن]. سقط من أ: للسداد والرشاد. زيادة في ر: وإخواننا حيث كانوا في البلاد

(٩) زيادة في ف: التي فيها

(١٠) ونبيّن حدودها: وحدودها [خ، غ]

كلّ إقليم من المدن الكبار والجبال والأنهار الطوال.[١] اعلم أيّها الأخ البارّ الرحيم أيّدك الله وإيّانا بروح منه بأنّ حدود الأقاليم تعتبر[٢] بساعات النهار وتفاوت الزيادة فيها وبيان ذلك أنّه إذا كانت الشمس في أوّل برج الحمل وإنّ طول النهار والليل وساعاتهم تتساوى[٣] في هذه الأقاليم كلّها فإذا سارت[٤] الشمس في درجات برج الحمل والثور والجوزاء اختلفت ساعات نهار كلّ إقليم[٥] حتّى إذا بلغت[٦] آخر الجوزاء الذي هو أوّل السرطان صار طول النهار في وسط الإقليم الأوّل ثلاث عشرة ساعة وفي وسط الإقليم الثاني ثلاث عشرة ساعة[٧] ونصف وفي وسط الإقليم الثالث أربع (١٦٩) عشرة ساعة وفي وسط (أ ٣٧) الإقليم الرابع أربع عشرة ساعة ونصف وفي وسط الإقليم الخامس خمس عشرة ساعة وفي وسط الإقليم السادس خمس عشرة ساعة ونصف[٨] وفي وسط الإقليم السابع ستّ عشرة ساعة سواء[٩] وفي المواضع التي عرضها ستّ وستّون درجة وما زاد[١٠] إلى

(١) سقط من أ: وإذ قد فرغنا من ذكر الأرض ووصفها وربعها المسكون فنريد أن نذكر الأقاليم السبعة ونبيّن حدودها طولا وعرضا وما في كلّ إقليم من المدن الكبار والجبال والأنهار الطوال

(٢) تعتبر: معتبرة [غ]، معتبرات [ف]، معتبر [ك، ل]

(٣) تتساوى: تتساوا [أ، خ]، متساو [ج، ز، ع، ل]، يتساويان [ط، ف، ن]

(٤) سارت: سوت [ع]

(٥) ساعات نهار كلّ إقليم: ساعات كلّ إقليم [ع]

(٦) زيادة في ن، ل: الشمس

(٧) سقط من ك: ساعة وفي وسط الإقليم الثاني ثلاث عشرة ساعة

(٨) سقط من خ: ونصف

(٩) سقط من ع: سواء

(١٠) زيادة في ع: على ذلك

التسعين درجة يصير نهاراً كلّه[١] وشرح كيفيّتها طويل[٢] مذكور في[٣] المجسطي.

واعلم أنّ معنى[٤] طول كلّ بلد ومدينة هو[٥] بعدها من أقصى المغرب ومعنى عرضها هو بعدها من[٦] خطّ الاستواء وخطّ الاستواء هو المواضع التي تكون الليل والنهار هناك أبداً متساويين فكلّ مدينة على ذلك الخطّ فلا عرض لها وكلّ مدينة في أقصى المغرب فلا طول لها أيضاً ومن أقصى المغرب إلى أقصى المشرق[٧] مائة وثمانون درجة مقدار كلّ درجة[٨] تسعة عشر فرسخاً[٩] وكلّ مدينة طولها تسعون درجة فهي في وسط من المشرق والمغرب وما كان أكثر فهي إلى المشرق أقرب وما كان أقلّ فهي[١٠] إلى المغرب أقرب.[١١] وكلّ مدينتين إحداهما أكثر طولاً وعرضاً فهي إلى المشرق والشمال أقرب من الأخرى[١٢]

---

(١) سقط من غ: يصير نهارا كلّه
(٢) كيفيّتها طويل: كيفيّتها كما هو [ع]، هذا يطول وهو [ف]
(٣) زيادة في ع: كتاب
(٤) سقط من غ: معنى. زيادة في ف: قولنا
(٥) هو: معنى [خ]
(٦) سقط من ط: من أقصى المغرب ومعنى عرضها هو بعدها من
(٧) ومن أقصى المغرب إلى أقصى المشرق: ومن أقصى المشرق إلى أقصى المغرب [ل]
(٨) سقط من غ: مقدار كلّ درجة
(٩) مقدار كلّ درجة تسعة عشر فرسخا: مقدار كلّ خمسة عشر درجة ساعة وهي تسعة عشر فرسخا [ف]
(١٠) سقط من غ: إلى المشرق أقرب وما كان أقلّ فهي
(١١) أقرب: أقلّ [ج]
(١٢) زيادة في غ: والتي بها قلّ طولا وعرضا فهي إلى المغرب والجنوب اقرب من الأخرى

والتفاوت الذي يكون بينهما في العرض كلّ درجة تسعة عشر فرسخاً[١] بالتقريب وأمّا تفاوتها في الطول[٢] فمختلف فما كان منها على خطّ الاستواء وكلّ درجة في الطول[٣] تسعة عشر فرسخاً[٤] وما كانت في الإقليم الأوّل فكلّ درجة[٥] سبعة عشر فرسخاً[٦] وفي الثاني كلّ درجة خمسة عشر فرسخاً[٧] وفي الثالث كلّ درجة ثلاثة عشر فرسخاً وفي الرابع كلّ درجة عشرة فراسخ[٨] وفي الخامس كلّ درجة سبعة فراسخ[٩] وفي السادس كلّ درجة خمسة فراسخ[١٠] وفي السابع كلّ درجة ثلاثة فراسخ.[١١]

---

(١) زيادة في ع: ١٩
(٢) زيادة في خ: والأرض
(٣) سقط من خ: فمختلف فما كان منها على خطّ الاستواء وكلّ درجة في الطول
(٤) تسعة عشر فرسخا: ١٩ فرسخا [ع]
(٥) زيادة في ف: في الطول
(٦) سبعة عشر فرسخا: ١٧ فرسخا [ع]، تسعة عشر فرسخا [غ]
(٧) خمسة عشر فرسخا: عشرة فراسخ ١٠ [ع]
(٨) سقط من ع: وفي الثالث كلّ درجة ثلاثة عشر فرسخا وفي الرابع كلّ درجة عشرة فراسخ
(٩) سبعة فراسخ: ٧ فراسخ [ع]
(١٠) خمسة فراسخ: ٥ فراسخ [ع]
(١١) ثلاثة فراسخ: ٣ فراسخ [ع]. زيادة في خ: والله أعلم بالصواب وإليه المرجع والمآب

ذكر أسماء المدن الكبار التي تنسب[١] في الأقاليم[٢] وهي كلّ مدينة عرضها أقلّ من اثنتي عشرة درجة ممّا يلي خطّ الاستواء أوّلها ممّا يلي المشرق.[٣]

(١٧٠)

أسماء المدن	الطول	العرض
قشمير من الهند		
مدينة الطيب من الهند		د ل
حضرموت من اليمن	عا ♉	يب[٤] ♉
وغارة[٥] من الحبشة		
كوكو من الحبشة		هـ ♉
مدينة على البحر العربي		

الإقليم الأوّل لزحل وطوله من المشرق إلى المغرب ٩٠٠٠ ميلاً[٦] ٣٠٠٠ فرسخاً[٧] وعرضه من الجنوب إلى الشمال ٤٤٠ ميلاً

(١) تنسب: نسبت [غ]، ليست [ز، ط، ع، ك، ن]. سقط من ف: تنسب

(٢) زيادة في ط، ع، غ، ف، ك، ن: السبعة

(٣) زيادة في ك: أسماء المدن قسمين من الهند مكر يا، جزيرة كول من السند قب ط، جزيرة الطيب من السند عب هـ من الحبش ح يا

(٤) يب: بد [أ]

(٥) وغارة: زغاوة [أ]

(٦) ٩٠٠٠ ميلا: ٦٠٠٠ ميلا [ج] ٥١٠٠٠ ميلا [خ]

(٧) ٣٠٠٠ فرسخا: ٣٠٠٠٠ فرسخا [غ]. زيادة في هامش ن: تسعة آلاف ميلا فيكون ثلاثة ألف. سقط من ع: ٣٠٠٠ فرسخا

١٤٢[١] فرسخاً[٢] وحدّه الأوّل ممّا يلي خطّ الاستواء حيث يكون ارتفاع القطب الشماليّ ثلاث عشرة درجة غير ربع[٣] وساعات نهاره الأطول[٤] اثنتا عشرة ساعة ونصف وربع[٥] ووسطه حيث يكون ارتفاع القطب عن الأفق ستّ عشرة درجة وثلثي[٦] درجة وساعات نهاره الأطول[٧] ثلاث عشرة ساعة[٨] وحدّه الثاني حيث يكون ارتفاع القطب عشرين[٩] درجة[١٠] ونصفاً وطول نهاره الأطول[١١] ثلاث عشرة[١٢] ساعة[١٣] وربع.

وفي هذا الإقليم من الجبال الطوال نحو من عشرين جبلاً منها ما طوله من عشرين فرسخاً إلى مائة فرسخ[١٤] إلى ألف

---

(١) ١٤٢: ١٤٦ [ج، ز، ط، ك، ن]، ٤١ [غ]

(٢) سقط من ع: ميلا ١٤٢ فرسخا. زيادة في ج: ٢ فرسخ. زيادة في هامش ن: أربعمائة وأربعين ميلا فيكون مائة وستّ وأربعة

(٣) غير ربع: وربع [أ]

(٤) الأطول: الأوّل [ع]

(٥) وربع: ربع [ع]

(٦) وثلثي: وثلثين [غ]، وثلاثا [ل]

(٧) الأطول: الأوّل [ع]

(٨) ثلاثة عشر ساعة: ١٣ ساعة [ع]، ثلاثة عشر ساعة وربع [خ]

(٩) عشرين: ٢٠ [ع]

(١٠) سقط من ع: درجة

(١١) الأطول: الأوّل [ع]

(١٢) ثلاث عشرة: ١٣ [ع]

(١٣) سقط من خ: وحدّه الثاني حيث يكون ارتفاع القطب عشرين درجة ونصفا وطول نهاره الأطول ثلاث عشرة ساعة

(١٤) سقط من ف: الطوال نحو من عشرين جبلا منها ما طوله من عشرين فرسخا إلى مائة فرسخ

فرسخ[١] وفيه أيضاً مقدار ثلاثين نهراً طوالاً منها ما طوله من عشرين فرسخاً إلى مائة فرسخ إلى ألف فرسخ وفيه[٢] من المدن المعروفة الكبار نحو من خمسين مدينة وابتداء هذا الإقليم من المشرق عن شمال جزيرة الياقوت[٣] فيمرّ على بلاد الصين ممّا يلي الجنوب ثمّ يمرّ على شمال بلد سرنديب ثمّ يمرّ على وسط بلاد الهند ثمّ يمرّ على وسط[٤] بلاد السند[٥] ثمّ يقطع بحر فارس ممّا يلي جنوب[٦] بلاد عمان[٧] ثمّ يمرّ على وسط بلاد الحبشة ويقطع نيل مصر (١٧١) هناك ثمّ يمرّ على بلاد النوبة ثمّ يمرّ على وسط بلاد البربر وبلاد ليبواني[٨] ثمّ يمرّ على جنوب بلاد مرطانة[٩] وينتهي إلى بحر المغرب وعامّة أهل هذه البلدان سودان.

---

(١) سقط من ع: منها ما طوله من عشرين فرسخا إلى مائة فرسخ إلى ألف فرسخ. سقط من غ: إلى ألف فرسخ

(٢) سقط من أ: مقدار ثلاثين نهرا طوالا منها ما طوله من عشرين فرسخا إلى مائة فرسخ إلى ألف فرسخ وفيه. سقط من ع: طوالا منها ما طوله من عشرين فرسخا إلى مائة فرسخ إلى ألف فرسخ وفيه

(٣) سقط من خ: الياقوت. سقط من ف: الكبار نحو من خمسين مدينة وابتداء هذا الإقليم من المشرق عن شمال جزيرة الياقوت

(٤) سقط من ف: شمال بلد سرنديب ثمّ يمرّ على وسط بلاد الهند ثمّ يمرّ على وسط

(٥) سقط من غ: ثمّ يمرّ على وسط بلاد السند

(٦) سقط من ف: ممّا يلي جنوب

(٧) عمان: اليمن [ف]. زيادة في ط، ك، ل: ثمّ يمرّ على وسط بلاد الشجر ثمّ يمرّ على وسط بلاد اليمن ثمّ يقطع بحر القلزم هناك. زيادة في ف: ثمّ يقطع بحر القلزم هناك

(٨) ليبواني: ليوناني [أ، ز]، سنوي [خ]، اليواني [ج]، كيوناي [ر]، البرابر [ن]، اليواي مع التصحيح «اليونان» [ط]، ينوني [ف]، البواي [ك]، النواني [ل]

(٩) ثمّ يمرّ على وسط بلاد الحبشة ويقطع نيل مصر هناك ثمّ يمرّ على بلاد النوبة ثمّ يمرّ على وسط بلاد البربر وبلاد ليبواني ثمّ يمرّ على جنوب بلاد مرطانة: ثم يمرّ على وسط بلاد السخر ثمّ يمرّ على [غ]

ذكر أسماء المدن التي في هذا الإقليم وهي كلّ مدينة عرضها من ثلاث عشرة درجة إلى عشرين درجة أوّلها ممّا يلي المشرق.

أسماء المدن	الطول	العرض
مدينة في أقصى الصين		
مدينة في جزائر الصين		
بايوقا من الهند	قل ♉	يب ♉
عدن من اليمن	سه ل	يج ♉
ديقلة من بلاد النوبة	نج م	يد ♉
مملكة الحبشة		
بلاد البربر		

الإقليم الثاني للمشتري وطوله من المشرق إلى المغرب ٩٦٠٠[١] ميلاً[٢] (١٧٢) وعرضه من الجنوب إلى الشمال ٤٠٠[٣] ميلاً وحدّه الأوّل ممّا يلي إقليم زحل حيث يكون ارتفاع القطب (ب ٣٧) عشرين درجة ونصفاً وطول نهاره الأطول ثلاث عشرة[٤] ساعة وربع[٥] ووسطه حيث يكون ارتفاع القطب[٦] كد درجة وستّ دقائق[٧] ونهاره الأطول ثلاث عشرة[٨] ساعة ونصف[٩] وحدّه الثاني[١٠] حيث يكون ارتفاع القطب عن الأفق كز[١١] درجة ونصف[١٢] ونهاره الأطول ثلاث عشرة ساعة ونصف وربع.

وفي هذا الإقليم من الجبال الطوال نحو من سبعة جبلاً[١٣]

---

(١) ٩٦٠٠: ١٦٠٠ [ج]، ٨٢٠٠ [ر، ن]، ٧٦٠٠ [خ، ز، ط، ك، ل]، ٩٦٨٦ [ع]

(٢) زيادة في هامش ن: ثمان الف وستّمائة وسبع وسبعون ميلا ٨٦٧٧

(٣) ٤٠٠: ٦٠٠ [ز، ك]. زيادة في هامش ن: أربعمائة ٤٠٠

(٤) ثلاث عشرة: يد عشر [أ]، ١٣ [ج]

(٥) سقط من ع: وحدّه الأوّل ممّا يلي إقليم زحل حيث يكون ارتفاع القطب عشرين درجة ونصفا وطول نهاره الأطول ثلاث عشرة ساعة وربع

(٦) زيادة في خ: عن الأفق

(٧) كد درجة وستّ دقائق: عشرين درجة و ستّ دقائق [ر]، ٢٤ درجة ٦ دقائق [ع]، أربعا وعشرين درجة وستّ دقائق [غ، ن]، أربعة وعشرين درجة وستّ دقائق [خ، ز، ط، ف، ك، ل]

(٨) ثلاث عشرة: ١٣ [ع]

(٩) زيادة في خ: وربع

(١٠) سقط من خ: الثاني

(١١) كز: سبعة وعشرين [خ، ر، ز، ط، غ، ف، ك، ل، ن]، ٢٧ [ع]

(١٢) سقط من خ، ع: ونصف

(١٣) نحو من سبعة جبلا: نحو من سبعة عشر [ج، ز، ط، ف، ك، ن]، ٢٧ [ع]، نحو من ستّ عشر [ل]

ومن الأنهار الطوال مثل ذلك[١] ومن المدن المعروفة الكبار نحو من خمسين مدنية وابتداء هذا الإقليم من المشرق إلى المغرب ويمرّ على وسط بلاد الصين ثمّ يمرّ على شمال بلاد سرنديب[٢] ثمّ يمرّ على بلاد الهند[٣] ممّا يلي الشمال[٤] ثمّ يمرّ على وسط بلاد الكابل ثمّ يمرّ على بلاد القندهار[٥] ثمّ يمرّ على شمال بلاد السند وجنوب بلاد مكران[٦] ثمّ يقطع بحر فارس[٧] ويمرّ على وسط بلاد العرب[٨] ثمّ يقطع بحر القلزم ويمرّ على شمال بلاد الحبشة وجنوب بلاد صعيد مصر[٩] ويقطع نيل مصر هناك ثمّ يمرّ على وسط بلاد إفريقية[١٠] ثمّ يمرّ على شمال[١١] بلاد البربر[١٢]

---

(١) سقط من ع: ومن الأنهار الطوال مثل ذلك

(٢) سرنديب: القندهار [غ]. سقط من ع: على وسط بلاد الصين ثمّ يمرّ على شمال بلاد سرنديب

(٣) سقط من ز، ع، ك: ثمّ يمرّ على بلاد الهند

(٤) سقط من خ: ثمّ يمرّ على شمال بلاد سرنديب ثمّ يمرّ على بلاد الهند ممّا يلي الشمال. سقط من ع، ل: ممّا يلي الشمال. زيادة في ز، ك، هامش ل: ثمّ يمرّ على بلاد القندهار. زيادة في ع: وبلاد القندهار

(٥) القندهار: الهند [أ]. سقط من غ: ثمّ يمرّ على بلاد الهند ممّا يلي الشمال ثمّ يمرّ على وسط بلاد الكابل ثمّ يمرّ على بلاد القندهار

(٦) مكران: كرمان [ط، ع، ف]

(٧) زيادة في ز، ع، ك: ويمرّ على بلاد عمان

(٨) العرب: الغرب [أ، ل]

(٩) صعيد مصر: الصعيد [ج]

(١٠) بلاد إفريقية: رقّة وإفريقية [ع، ف]، الرقّة وإفريقية [ك، ل]

(١١) شمال: جنوب [ع]. سقط من ج: شمال

(١٢) سقط من خ: وجنوب بلاد صعيد مصر ويقطع نيل مصر هناك ثمّ يمرّ على وسط بلاد إفريقية ثمّ يمرّ على شمال بلاد البربر

وجنوب(١) بلاد القيروان ثمّ يمرّ على وسط بلاد مرطانة وينتهي إلى بحر المغرب وأكثر أهل هذه البلاد ألوانهم ما بين السمرة إلى السواد.

ذكر المدن الكبار التي في هذا الإقليم أوّلها ممّا يلي المشرق في أقصى بلاد الصين وهي كلّ مدينة عرضها من ك(٢) إلى كز ل(٣) وأوّلها ممّا يلي المشرق.

أسماء المدن	الطول	العرض
طوعما من الصين	قمب م	ك مه
طعولا من الصين	قمح ى	كد ل
المنصورة(٤) من السند	صح ♉	كب ♉

(١) وجنوب: وعلى شمال [ع]

(٢) ك: عشرين درجة [ر، غ]

(٣) كز ل: سبع وعشرين درجة وثلاثين دقيقة [ر]، سبع وعشرين درجة وثلاثين [غ]، ك درجة ودقيقة [ف]، كز درجة ول دقيقة [ن]

(٤) المنصورة: المقصورة [أ]

الطائف من اليمن	سز ى	كا ♉
مكّة من التهامة	سز ♉	كا ى
يثرب مدينة الرسول	سه ♉	كه ♉
أخميم صعيد مصر	نه ل	كز ك
إفريقية من الغرب		

الإقليم الثالث للمرّيخ وطوله من المشرق إلى المغرب ٩٢٣٦ ميل(١) وعرضه من الجنوب إلى الشمال ٣٥٠ ميل(٢) وحدّه من سبع وعشرين درجة ونصف إلى ثلاث وثلاثين درجة وثلاثين دقيقة(٣) ووسطه حيث يكون ارتفاع القطب عن الأفق ثلاثين(٤) درجة ونصفاً وخمساً(٥) نهاره الأطول أربع عشرة ساعة(٦) سواء.

وفي هذا الإقليم من الجبال الطوال ثلاثة وثلاثون(٧) جبلاً

(١) ٩٢٣٦ ميل: ٦٠٠ [ج]، ٢٠٠ [خ]، ٦٢٥٥ ميل [ر]، ٨٢٠٠ ميلا [ز، ط، غ، ف، ك، ل، ن]، ٢٠٠٠ ميلا [ع]. زيادة في هامش ن: ثمان ألوف ومائتي وسبع وسبعين ٨٢٧٧

(٢) ٣٥٠ ميل: ٣٠٠ ميلا [ع]، ٣٥٠٠ ميلا [ف]. زيادة في هامش ن: ثلاثمائة وخمسين ميلا

(٣) ثلاث وثلاثين درجة وثلاثين دقيقة: ثلاثين درجة وثلاثين دقيقة [أ]، لج د ول دقيقة [ج]. وحدّه الأوّل من سبعة وعشرين درجة ونصف إلى ثلاث وثلاثين درجة وثلاثين دقيقة: وحدّه الأوّل من آخر إقليم المشتري [ع]

(٤) ثلاثين: ثلثي [خ]

(٥) وخمساً: وخمسين دقيقة [ج]

(٦) أربع عشرة ساعة: أربعة وعشرون ساعة [ز]

(٧) وثلاثون: وثلاثين [أ، ج، خ]

ومن الأنهار الطوال اثنان وعشرون[1] نهراً ومن البلدان[2] المعروفة الكبار[3] مائة وثمانية[4] وعشرون[5] مدينة[6] وابتداء هذا الإقليم من المشرق فيمرّ على شمال بلاد الصين وجنوب[7] بلاد ياجوج وماجوج[8] ثمّ يمرّ على شمال[9] بلاد الهند[10] وجنوب بلاد الترك[11] ثمّ يمرّ على وسط[12] بلاد كابل[13] ثمّ يمرّ[14] على بلاد القندهار[15] ثمّ على[16] مكران ثمّ يمرّ[17] على جنوب[18] بلاد سجستان[19] ثمّ يمرّ على وسط بلاد[20] كرمان[21] ثمّ يمرّ[22]

(١) اثنان وعشرون: اثنان وعشرين [خ]، ثلاثة وعشرون [ف]، اثني وعشرون [ن]
(٢) البلدان: المدينة [ع]، المدن [ز، ف، ك، ن]
(٣) سقط من خ، ع: الكبار
(٤) وثمانية: مدينة [خ]، وثمان [ر، ع، غ، ن]، ثمنية [ط، ف]
(٥) وعشرون: وعشرين [ر، ف، ن]
(٦) زيادة في ع: كبارا
(٧) وجنوب: في جنوب [ع]
(٨) سقط من ج، ز، ط، ع، ن: وماجوج
(٩) سقط من ط، ع: شمال
(١٠) سقط من خ: وجنوب بلاد ياجوج وماجوج ثمّ يمرّ على شمال بلاد الهند
(١١) الترك: سجستان [غ]
(١٢) سقط من ل: وسط
(١٣) سقط من ع: ثمّ يمرّ على وسط بلاد كابل
(١٤) سقط من ط: يمرّ
(١٥) القندهار: قندهار [أ، ج، ط، ع، ك، ل]. زيادة في ع: ثمّ يمرّ على بلاد الهند. سقط من غ: ثمّ يمرّ على وسط بلاد كابل ثمّ يمرّ على بلاد القندهار
(١٦) ثمّ على: ثمّ يمرّ بلاد على [ر]، ثمّ على بلاد [ع، ن]
(١٧) سقط من ر، ع، ك، ل، ن: يمرّ
(١٨) سقط من ع: جنوب
(١٩) جنوب بلاد سجستان: جنوب بلاد سجستان جنوبا [ج]
(٢٠) سقط من ع: بلاد

على بلاد (١٧٤) فارس ممّا يلي البحر[١] ثمّ يمرّ على بلاد[٢] العراق ممّا يلي الجنوب ثمّ يمرّ على جنوب[٣] بلاد[٤] ديار[٥] بكر وشمال ديار[٦] العرب ثمّ يمرّ على وسط بلاد الشام[٧] ثمّ يمرّ على بلاد مصر ثمّ على بلاد الإسكندريّة ثمّ على وسط بلاد مرماريقى[٨] ثمّ على وسط بلاد القادسيّة[٩] ثمّ على وسط بلاد القيروان ثمّ على بلاد طنجة[١٠] ثمّ ينتهي إلى بحر المغرب وأكثر أهل هذه البلدان سمر.

---

(٢١) سقط من خ: ثمّ يمرّ على بلاد القندهار ثمّ على مكران ثمّ يمرّ على جنوب بلاد سجستان ثمّ يمرّ على وسط بلاد كرمان

(٢٢) سقط من خ: يمرّ

(١) سقط من ف: ثمّ يمرّ على بلاد القندهار ثمّ على مكران ثمّ يمرّ على جنوب بلاد سجستان ثمّ يمرّ على وسط بلاد كرمان ثمّ يمرّ على بلاد فارس ممّا يلي البحر

(٢) زيادة في خ: بحر

(٣) سقط من ع: جنوب

(٤) سقط من خ: بلاد

(٥) ديار: داير [خ]. سقط من ك: ديار

(٦) ديار: بلاد [خ، ر، غ، ك، ل، ن]

(٧) ثمّ يمرّ على وسط بلاد الشام: ثمّ على وسط بلاد الشاتي [خ]، ثمّ على وسط الشمال [ك]

(٨) وسط بلاد مرماريقى: وسط برماريقى [ع]

(٩) القادسيّة: الغاستيّة [أ]، القاسية [ز، ط، ك، ل]. سقط من خ: ثمّ على وسط بلاد مرماريقى ثمّ على وسط بلاد القادسيّة

(١٠) طنجة: تنجة [ن]

ذكر المدن التي في هذا الإقليم وهي كلّ مدينة عرضها من كز ل إلى لجـ م[١] أوّلها ممّا يلي المشرق.

أسماء المدن	الطول	العرض
القندهار من الهند	قي ♉	ل ♉
كابل من الهند	ق ♉	لح ♉
السرجان من كرمان	قح ♉	لب ل
شيراز من فارس	عح ♉	ل ♉
أهواز من خوزستان	عه ♉	لب ♉
البصرة[٢] من العراق	عد ♉	لا ♉
الكوفة من العراق	سط ♉	لب ♉
دمشق من الشام	س ♉	لحـ ♉
بيت المقدّس من فلسطين	نو ♉	لحـ ♉
الفسطاط من مصر	له ♉	كط ♉
الإسكندريّة	نط ♉	لا ♉
القيروان من الغرب	لا ♉	لب ♉

(١) من كز ل إلى لجـ م: من كز درجة ول دقيقة إلى لج درجة وم دقيقة [ن]، من كز ك إلى لج د [ل]، من سبع وعشرين درجة وثلاثين دقيقة إلى ثلاث وثلاثين درجة وثلاثين دقيقة [ر]. سقط من ع، غ، ف: وهي كل مدينة عرضها من كز ل إلى لجـ م

(٢) البصرة: المنصورة [أ]

(١٧٥)

الإقليم الرابع للشمس وطوله من المشرق إلى المغرب ٤٠٠٠ ميلاً[١] وعرضه من الجنوب إلى الشمال ٣٠٠[٢] ميلاً[٣] وحدّه من ثلاث وثلاثين درجة[٤] وثلاثين دقيقة[٥] إلى تسع وثلاثين درجة[٦] ووسطه حيث يكون ارتفاع القطب عن الأفق ستّاً وثلاثين[٧] درجة وخمسين دقيقة[٨] ونهاره الأطول أربع عشرة ساعة[٩] ونصف وفي هذا الإقليم من الجبال الطوال خمسة وعشرون جبلاً ومن الأنهار الطوال اثنان وعشرون نهراً[١٠] ومن المدن المعروفة الكبار نحو من مائتين[١١] واثنتي عشرة[١٢] مدينة وابتداء هذا الإقليم من المشرق ويمرّ على شمال بلاد الصين وجنوب بلاد ياجوج ثمّ يمرّ

(١) ٤٠٠٠ ميلا: ٣٠٠ ميلا [خ]، ٧٨٠٠ ميلا [ز، ط، ك]، ٨٠٠ ميلا [ع]. زيادة في هامش ن: سبعة ألوف وتسعمائة وسبعة ميلا ٧٩٠٧ . سقط من ج، ر، ن: ٤٠٠٠ ميلا

(٢) ٣٠٠: ٤٦٦ [خ]

(٣) زيادة في هامش ن: ثلاثمائة وسبعين وسبعة ميلا ٣٠٧٧

(٤) من ثلاث وثلاثين درجة: من ثلاثين درجة [ن]

(٥) سقط من ع: وثلاثين دقيقة

(٦) سقط من ن: وثلاثين دقيقة إلى تسع وثلاثين درجة

(٧) ستّا وثلاثين: ٣٦ [ج، ع]

(٨) وخمسين دقيقة: ن دقيقة [ج]، وكسر [ع]

(٩) أربع عشرة ساعة: يد ساعة [ج]

(١٠) اثنان وعشرون نهرا: خمسة وعشرون نهرا [خ، ف]

(١١) مائتين: مائتي [ل]

(١٢) مائتين واثنتي عشرة: مائتين واثنان عشر [أ]، مائتي واثني عشر [ج، ط، ك]، مائتي واثنا عشر [خ، ف]، مائتين واثنتا عشرة [ع]، مائتين مدينة واثني عشرة [غ]

على بلاد الترك ممّا يلي الجنوب ثمّ إلى بلاد الهند[1] ثمّ يمرّ على[2] شمال بلاد[3] بلخ[4] وباميان[5] ثمّ يمرّ على شمال[6] بلاد مكران[7] ثمّ يمرّ على وسط بلاد[8] سجستان ثمّ على وسط بلاد فارس وخوزستان[9] ثمّ يمرّ على وسط بلاد العراق ثمّ يمرّ على وسط بلاد ديار ربيعة وديار بكر[10] ثمّ يمرّ على جنوب بلاد الثغر وشمال بلاد الشام ويمرّ على وسط[11] بحر[12] الروم وجزيرة قبروس ويمرّ في البحر على شمال بلاد مصر والإسكندريّة[13]

---

(١) ثمّ إلى بلاد الهند: وشمال بلاد الهند [ج، ط، غ، ل]. زيادة في ز: وتخارستان. زيادة في ع: ثمّ يمرّ على بلاد تخارستان. زيادة في ك: وطخارستان. سقط من ع: ممّا يلي الجنوب ثمّ إلى بلاد الهند.

(٢) سقط من خ: ثمّ إلى بلاد الهند ثمّ يمرّ على

(٣) زيادة في خ: العرب

(٤) بلخ: بلخ خراسان [خ]. ثمّ يمرّ على شمال بلاد بلخ: وبلج [ع]

(٥) سقط من غ: ثمّ يمرّ على شمال بلاد بلخ وباميان.

(٦) سقط من ع: شمال

(٧) مكران: كابلان [أ، ج، غ] ايلان [خ]. سقط من ر: ثمّ يمرّ على شمال بلاد مكران

(٨) سقط من ع، ن: وسط بلاد

(٩) سقط من ج، خ: ثمّ على وسط بلاد فارس وخوزستان. سقط من غ: ثمّ يمرّ على وسط بلاد سجستان ثمّ على وسط بلاد فارس وخوزستان

(١٠) سقط من ع: وديار بكر. سقط من ن: ثمّ يمرّ على وسط بلاد ديار ربيعة وديار بكر

(١١) ويمرّ على وسط: ثمّ يقطع [ع]

(١٢) بحر: بلاد [خ]

(١٣) زيادة في ز، ع، ك، هامش ل: ثمّ على جزيرة سقليّة. سقط من ع: ويمرّ في البحر على شمال بلاد مصر والإسكندريّة

وشمال بلاد مرماريقى وبلاد (أ ٣٨) قادسيّة[١] وبلاد القيروان وبلاد طنجة وينتهي إلى بحر المغرب وأكثر أهل هذه البلدان ألوانهم ما بين السمرة[٢] والبياض.

وهذا الإقليم هو إقليم[٣] الأنبياء والحكماء لأنّه وسط الأقاليم ثلاثة منها جنوبيّة وثلاثة شماليّة وهو أيضاً في قسمة الشمس النيّر الأعظم وأهل هذا الإقليم أعدل الناس طباعاً وأخلاقاً[٤] ثمّ بعده أهل[٥] الإقليمين اللذين[٦] عن جنبيه[٧] أعني[٨] الثالث والخامس وأمّا أهل الأقاليم الباقية فناقصون[٩] عن طبيعة الأفضل لأنّ صورتهم سمجة وأخلاقهم وحشة[١٠] مثل الزنج والحبشة وأكثر الأمم[١١] الذين هم في الإقليم الأوّل والثاني وكذلك الأمم[١٢]

---

(١) قادسيّة: الفاشة [خ]، قاسيّة [ز، ل]، القاشية [ط]، القادسيّة [ف]

(٢) السمرة: السواد [خ]

(٣) هو إقليم: معدن [ع]

(٤) سقط من أ، ج، خ، ر، ط، غ، ف: وأهل هذا الإقليم أعدل الناس طباعا وأخلاقا

(٥) سقط من خ، غ، ف، ك: أهل

(٦) أهل الإقليمين اللذين: الإقليمان اللذان [ع، ن]

(٧) عن جنبيه: عن جنبته [خ]، جنبه [غ]

(٨) سقط من ع: أعني. زيادة في ف: الإقليم

(٩) أهل الأقاليم الباقية فناقصون: الأقاليم الباقية فأهلها ناقصون [ج، خ، ز، ن، غ، ف، ك، ل]

(١٠) لأنّ صورتهم سمجة وأخلاقهم وحشة: والدليل على ذلك سماجة صورهم وسوء أخلاقهم [ع]

(١١) سقط من ع: وأكثر الأمم. سقط من ف: لأنّ صورتهم سمجة وأخلاقهم وحشة مثل الزنج والحبشة وأكثر الأمم

(١٢) سقط من غ: الأمم

الذين هم في الإقليم السادس والسابع مثل ياجوج وماجوج والبلغر(١) والصقالبة(٢) وأمثالهم.(٣)

أسماء المدن التي في هذا الإقليم وهي كلّ مدينة عرضها من لجـ م إلى لط.(٤)

(١٧٦)

أسماء المدن	الطول	العرض
سمرقند وراء النهر	فح ك	م ♉
بلخ من خراسان	فح له	لح م
هراة من خراسان		
مرو من خراسان		لح ن
نيسابور من خراسان	ف مه	لز ♉

(١) والبلغر: واليغر [أ، ع، ل]، والياغر [ر]، الثغر [ز]، والتلغر [خ]، والتغزغز [ط]

(٢) والصقالبة: والثقالبة [خ]

(٣) والبلغر والصقالبة وأمثالهم: والصين الذين هم في الإقليم السابع [ع]

(٤) من لجـ م إلى لط: من ثلاث وثلاثين درجة إلى تسع وثلاثين درجة [ر، غ]، من لج م إلى لح [ف]. أسماء المدن التي في هذا الإقليم وهي كلّ مدينة عرضه من لجـ م إلى لط: وهي كل مدينة عرضه من لج م إلى لط والله تعالى أعلم أسماء المدن التي في الإقليم الرابع [ز]، وهي كل مدينة عرضه من لج درجة وم دقيقة إلى لط درجة وأسماء المدن التي في هذا الإقليم [ن]

جرجان من الجبل		
آمل من طبرستان		
الريّ	عه	له مه
الديلم من جيلان		
إصفهان من فارس	فو ♉	لب كه
همذان من بلاد ماهان	عح ♉	لو ♉
بغداد من العراق	ع ♉	لح كه
الموصل من ديار ربيعة	سح ى	لو ل
حلب من الشام	سح ♉	له ♉
أنطاكية	سا لد	لد ي
عمّوريّة	نح ♉	م ♉
أندلس من المغرب		

الإقليم الخامس للزهرة طوله من المشرق إلى المغرب ٤٠٠٠ ميلاً[1] وعرضه من الجنوب إلى الشمال ٢٥٥ ميلاً[2] وحدّه من

---

(١) ٤٠٠٠ ميلا: ٤٠٠ ميلا [أ]، ٧٤٠٠ ميلا [ز، ط، ك، ل]، أربعة ألف ميلا [ف]. سقط من ج، خ، ر، غ، ن: ٤٠٠٠ ميلا. زيادة في هامش ن: أربعة ألوف ميل ٤٠٠٠

(٢) ٢٥٥ ميلا: ٢٤٥ ميلا [ر، غ، ك]، ٢٥٠ ميلا [ط، ل]، ٢٤٤ ميلا [ن]، ٢٥٥ ميلا [هامش ل]. زيادة في هامش ن: مائتي وخمسون ميلا ٢٥٠

تسع وثلاثين درجة إلى ثلث وأربعين درجة ونصف[١] ووسطه من حيث يكون ارتفاع القطب إحدى وأربعين درجة وثلثا ونهاره الأطول خمس عشرة ساعة سواء.

وفي هذا الإقليم من الجبال الطوال نحو من ثلاثين جبلاً ومن الأنهار الطوال نحو من خمسة عشر نهراً ومن المدن المعروفة الكبار نحو من مائتي (١٧٧) مدينة وابتداؤه من المشرق يمرّ على وسط بلاد ياجوج[٢] ثمّ يمرّ على وسط بلاد الترك ثمّ يمرّ على بلاد فرغانة وبلاد اسبيجاب[٣] ثمّ يمرّ على وسط بلاد السغد[٤] وما وراء النهر[٥] ثمّ يقطع جيحون ويمرّ على وسط[٦] بلاد[٧] خراسان ثمّ يمرّ على شمال سجستان وكرمان ثمّ يمرّ على شمال بلاد فارس ووسط بلاد الريّ والماهان[٨] ثمّ يمرّ على شمال[٩] بلاد[١٠] العراق وجنوب بلاد آذربيجان ثمّ يمرّ على وسط أرمينية

---

(١) من تسع وثلاثين درجة إلى ثلث وأربعين درجة ونصف: من لط إلى مج درجة ونصف [ج]. سقط من خ: ونصف

(٢) زيادة في ز، ع، ف، ك: وماجوج

(٣) اسبيجاب: سيحان [خ]. سقط من ع: وبلاد اسبيجاب

(٤) السغد: الصين [أ]، السعد [ج، ك، ل]، الصعيد [خ، ز]، الصعد [ط، ع، ف] السند [ن]

(٥) ثمّ يمرّ على وسط بلاد السغد وما وراء النهر: ثمّ على وسط بلاد ما وراء النهر [ع]

(٦) سقط من ع: وسط

(٧) سقط من غ: بلاد

(٨) سقط من غ: ثمّ يمرّ على شمال سجستان وكرمان ثمّ يمرّ على شمال بلاد فارس ووسط بلاد الريّ والماهان

(٩) شمال: وسط [خ]. سقط من ع: شمال

(١٠) سقط من ج، ن: ووسط بلاد الريّ والماهان ثمّ يمرّ على شمال بلاد

وشمال بلاد الثغر ثمّ على وسط بلاد الروم ويقطع خليج قسطنطينيّة هناك ويمرّ على شمال[1] بحر الروم[2] ووسط بلاد الرومية ويمرّ على جنوب[3] هيكل الزهرة[4] ثمّ يمرّ على وسط بلاد الأندلس[5] وينتهي إلى بحر المغرب وأكثر أهل هذه البلدان أبيض.[6]

أسماء المدن التي في هذا الإقليم وهي كلّ مدينة عرضها من لط[7] إلى مجـ ل.[8]

أسماء المدن	الطول	العرض
بلاد ياجوج وماجوج		
بلاد خاقان من الترك		
الطراز من الترك	فز ن	لط ل
خوارزم من ما وراء النهر		
أردبيل من آذربيجان		

(١) سقط من ف: شمال
(٢) سقط من ل: ويقطع خليج قسطنطينيّة هناك ويمرّ على شمال بحر الروم
(٣) سقط من ع: جنوب
(٤) زيادة في هامش ن: ثمّ على النيل
(٥) الأندلس: رومية والأندلس [خ]
(٦) أبيض: بيض [ج، خ، ز، ط، غ، ف، ك، ل]، بيض البشرة [ع، ن]
(٧) لط: تسع وثلاثين درجة [ر، غ]
(٨) مجـ ل: ثلاث وأربعين درجة وثلاثين دقيقة [غ]، ثلاث وأربعين درجة وثلاث دقيقة [ر]، مج درجة ول دقيقة [ن]، مر ل [ل]. سقط من ع: وهي كلّ مدينة عرضها من لط إلى مجـ ل

خلاط من أرمينية	سه ♉	لط م
ملطية من أرمينية	سا ♉	لط ♉
ماقيذونية[(١)]	مط ♉	م ♉
رومية الكبرى من الروم	له فى	ما ن
هيكل الزهرة	نط ل	مب ي

الإقليم السادس لعطارد وطوله من المشرق إلى المغرب ٥٠٠٠ ميلاً[(٢)] وعرضه من الجنوب إلى الشمال ٢١٦[(٣)] ميلاً وحدّه من ثلاث وأربعين درجة[(٤)] ونصف إلى سبع وأربعين درجة وربع[(٥)] ووسطه حيث يكون ارتفاع القطب (١٧٨) خمساً وأربعين[(٦)] درجة وخمسين دقيقة[(٧)] ونهاره الأطول خمس عشرة[(٨)] ساعة ونصف.[(٩)]

---

(١) ماقيذونية: ملطية من أرمينية [أ]

(٢) ٥٠٠٠ ميلا: ٧٠٠٠ ميلا [ز، ط، ل]، ٢٠٠ [ط]، ٧٠٠ ميلا [ك]. سقط من أ، ج، خ، ر، غ، ف، ن: ٥٠٠٠ ميلا

(٣) ٢١٦: ٢٠٠ [ز، ك]، ٢١٠ [ج، خ، ل، ن]، ٢٢٠ [ف]. زيادة في هامش ن: مائتي وعشرة. سقط من ع، غ: ٢١٦

(٤) ثلاث وأربعين درجة: مج درجة [ج]

(٥) سبع وأربعين درجة وربع: مز درجة وربع [ج]، خمسين درجة ونصف [ع]. سقط من ر: وعرضه من الجنوب إلى الشمال ٢١٦ ميلا وحدّه من ثلاث وأربعين درجة ونصف إلى سبع وأربعين درجة وربع

(٦) خمسا وأربعين: مه [ج]

(٧) وخمسين دقيقة: ن دقيقة [ج]، وثلثين [ع]، وخمس دقائق [ل]. وعرضه من الجنوب إلى الشمال ٢١٦ ميلا وحدّه من ثلاث وأربعين درجة ونصف إلى سبع وأربعين درجة وربع ووسطه حيث يكون ارتفاع القطب خمسا وأربعين درجة

وفي هذا الإقليم من الجبال الطوال نحو من اثنين وعشرين[١] جبلاً ومن الأنهار الطوال نحو من اثنين وثلاثين نهراً ومن المدن المعروفة الكبار تسعون[٢] مدينة وابتداؤه من المشرق[٣] فيمرّ على شمال[٤] بلاد ياجوج ثمّ يمرّ على جنوب[٥] بلاد سجستان[٦] ثمّ على جنوب بلاد التغزغز[٧] ثمّ يمرّ على وسط بلاد خاقان وجنوب بلاد كيماك[٨] ثمّ يمرّ على شمال بلاد السغد[٩] وما وراء النهر[١٠] ثمّ يمرّ على وسط بلاد خوارزم[١١] ثمّ يمرّ على شمال

وخمسين دقيقة: ووسطه حيث يكون ارتفاع القطب خمسة وأربعين درجة وخمسين دقيقة وعرضه من الجنوب إلى الشمال وحدّه من ثلاث وأربعين درجة ونصف إلى سبع وأربعين درجة وخمس عشرة دقيقة [ر]

(٨) خمس عشرة: ستّ عشرة [ع]

(٩) سقط من خ، ع: ونصف

(١) اثنين وعشرين: أحد عشر [ع]

(٢) تسعون: اثنتان وعشرون [ع]، نحو من سبعين [ر]

(٣) زيادة في خ: إلى المغرب

(٤) شمال: جنوب [ع]. سقط من غ: شمال

(٥) جنوب: شمال [ل]. سقط من ع: جنوب

(٦) سجستان: سيستان [ع، غ]. سقط من خ، ن: ثمّ يمرّ على جنوب بلاد سجستان

(٧) التغزغز: اليغر [أ]، الثغر [ج، خ، ز، ل، ن]، تغرغر [ط، ع، غ، ف]

(٨) كيماك: كيمال [أ، ط، ك]، كيلان [ج، ر] كيال [خ]، كيمان [ف]. ثمّ يمرّ على جنوب بلاد سجستان ثمّ على جنوب بلاد التغزغز ثمّ يمرّ على وسط بلاد خاقان وجنوب بلاد كيماك: ثمّ يمرّ على بلاد كيمال [غ]. زيادة في ك، ل: ثمّ يمرّ على شمال بلاد اسبيجاب

(٩) السغد: الصعيد [أ]، الصعد [ط، ف]

(١٠) زيادة في هامش ز: ثمّ يمرّ على وسط بلاد خاقان وجنوب بلاد كيمال ثمّ يمرّ على شمال بلاد السغد وما وراء النهر

(١١) سقط من ج: ثمّ يمرّ على شمال بلاد السغد وما وراء النهر ثمّ يمرّ على وسط بلاد خوارزم

بلاد إسبيجاب[١] ثمّ يمرّ على شمال بلاد جرجان (١٧٩) وطبرستان والديلم وجيلان[٢] ويقطع بحر[٣] طبرستان[٤] ويمرّ على وسط بلاد آذربيجان ثمّ يمرّ على وسط بلاد أرمينية وملطية ثمّ يمرّ على[٥] شمال بحر بنطس[٦] ثمّ يمرّ على شمال قسطنطينيّة ثمّ يمرّ على وسط بلاد[٧] ماقيذونية[٨] ثمّ يمرّ على وسط بلاد إفريقية وممّا يلي الشمال ويمرّ على جنوب بحر الصقالبة ثمّ يمرّ على شمال هيكل الزهرة[٩] وينتهي إلى بحر المغرب[١٠] وأكثر أهل

(١) ثمّ يمرّ على شمال بلاد السغد وما وراء النهر ثمّ يمرّ على وسط بلاد خوارزم ثمّ يمرّ على شمال بلاد إسبيجاب: ثمّ يمرّ على شمال اسحاب ثمّ يمرّ على شمال بلاد الصعيد وما وراء النهر ثمّ يمرّ على وسط بلاد خوارزم [خ]، ثم يمرّ على شمال بلاد إسبيجاب ثمّ يمرّ على شمال بلاد الصعيد وما وراء النهر ثمّ يمرّ على وسط بلاد خوارزم [ر]، ثمّ يمرّ على شمال بلاد إسبيجاب ثمّ يمرّ على شمال بلاد الصغد وما وراء النهر ثمّ يمرّ على وسط بلاد خوارزم [ط]، ثمّ يمرّ على شمال بلاد إسبيجاب ثمّ يمرّ على شمال بلاد السغد وما وراء النهر ثمّ يمرّ على وسط بلاد خوارزم [ن]. سقط من غ: ثمّ يمرّ على شمال بلاد السغد وما وراء النهر ثمّ يمرّ على وسط بلاد خوارزم. سقط من ز، ف، ل: ثمّ يمرّ على شمال بلاد إسبيجاب

(٢) وجيلان: وكيلان [أ]

(٣) بحر: بحيرة [غ]

(٤) سقط من ك: ثمّ يمرّ على شمال بلاد إسبيجاب ثمّ يمرّ على شمال بلاد جرجان وطبرستان والديلم وجيلان ويقطع بحر طبرستان

(٥) سقط من ف: على وسط بلاد آذربيجان ثمّ يمرّ على وسط بلاد أرمينية وملطية ثمّ يمرّ على

(٦) بنطس: فلسطين [ف]

(٧) وسط بلاد: شمال بحر [خ]

(٨) ماقيذونية: قيذونية [ز، ط، ك]، ماقدونة [غ]

(٩) ثمّ على جنوب بلاد التغزغز ثمّ يمرّ على وسط بلاد خاقان وجنوب بلاد كيماك ثمّ يمرّ على شمال بلاد السغد وما وراء النهر ثمّ يمرّ على وسط بلاد خوارزم

هذه البلدان ألوانهم ما بين الشقرة والبياض.[١]

أسماء المدن التي في هذا الإقليم وهي كلّ مدينة عرضها من مجـ ل[٢] إلى مز يه[٣] أوّلها ممّا يلي المشرق (ب ٣٨).

أسماء المدن	الطول	العرض
بلاد ياجوج	فز ♉	مب ♉
بلاد الخزر	صح ♉	مه ♉
جزيرة في جرجان		
برذعة من آذربجان	عح ♉	مجـ ♉
باب الأبواب		

ثمّ يمرّ على شمال بلاد إسبيجاب ثمّ يمرّ على شمال بلاد جرجان وطبرستان والديلم وجيلان ويقطع بحر طبرستان ويمرّ على وسط بلاد آذربيجان ثمّ يمرّ على وسط بلاد أرمينية وملطية ثمّ يمرّ على شمال بحر بنطس ثمّ يمرّ على شمال قسطنطينيّة ثمّ يمرّ على وسط بلاد ماقيذونية ثمّ يمرّ على وسط بلاد إفريقية وممّا يلي الشمال ويمرّ على جنوب بحر الصقالبة ثمّ يمرّ على شمال هيكل الزهرة: ثمّ بلاد تغرغر ثمّ بلاد كيمال ثمّ بلاد الآن وشمال بحر جرجان وشمال بلاد ماقدونة ثمّ يمرّ على جنوب بحر الصقالبة وجزيرة الريّ [ع]

(١٠) زيادة في ك: ثمّ يمرّ على شمال بلاد جرجان وطبرستان والديلم وكيلان ويقطع بحر طبرستان

(١) ما بين الشقرة والبياض: شقر [ع]

(٢) مجـ ل: يحرل [خ]، ثلاث وأربعين درجة وثلاثين دقيقة [ر، غ]، ٢ [ع]، مجـ بد [ك]. سقط من ع: مجـ ل

(٣) مز يه: مح [ج]، سبع وأربعين درجة وخمسة عشر دقيقة [ر]، مط [ع]، ستّ وأربعين درجة وخمس عشرة دقيقة [غ]

قسطنطينيّة الروم	مط ن	مه ♉
بلاد برجان	م ♉	مه ♉
بلاد افيليون	كه مه	مد نه

الإقليم السابع للقمر وطوله من المشرق إلى المغرب ٦٦٠٠ ميلاً[١] وعرضه من الجنوب إلى الشمال ١٨٥ ميلاً[٢] وحدّه من سبع وأربعين درجة[٣] وربع[٤] إلى خمسين درجة[٥] ونصف ووسطه حيث يكون[٦] ارتفاع القطب عن الأفق مح[٧] درجة وثلثي[٨] دقيقة[٩] فطول نهاره الأطول[١٠] يو[١١] ساعة سواء وفي

(١) سقط من أ، ج، خ، ر، غ ، ف، ن: ٦٦٠٠ ميلا. زيادة في هامش ن: ستّة ألوف وستمائة ميلا ٦٦٠٠. زيادة في هامش ل: ٦٦٠٠ ميلا

(٢) سقط من ط، ع: ١٨٥ ميلا. سقط من غ: ٦٦٠٠ ميلا وعرضه من الجنوب إلى الشمال ١٨٥ ميلا. زيادة في هامش ن: مائة وخمسة وثمانين

(٣) سبع وأربعين درجة: مز درجة [ج]

(٤) سقط من ع، غ: وربع

(٥) خمسين درجة: ن درجة [ج]، خمس درجة [ع]

(٦) سقط من ع: حيث يكون

(٧) مح: ثمانيا وأربعين [ر، غ]، ثمانية وأربعين [خ، ز، ط، ك]، ثمانية وأربعون [ل]، ثمان [ع]، ثمان وأربعين [ن]

(٨) وثلثي: وخمسين [ر]، وثلاثين [ع، ن]، وثلثا [ل]

(٩) سقط من أ، ج، خ، ز، ط، ع، ك: دقيقة

(١٠) سقط من ع: الأطول

(١١) يو: ستّ عشرة [خ، ر، ز، ط، ع، غ، ك، ل، ن]. سقط من ف: مح درجة وثلثي دقيقة فطول نهاره الأطول يو

هذا الإقليم(١) من الجبال الطوال نحو من عشرة أجبال(٢) ومن الأنهار الطوال نحو من أربعين نهراً ومن(٣) المدن المعروفة الكبار اثنان وعشرون مدينة وابتداؤه من المشرق ويمرّ على جنوب(٤) بلاد ياجوج وماجوج(٥) وبلاد سيستان(٦) ثمّ بلاد التغزغز(٧) ثمّ بلاد كيماك(٨) ثمّ يمرّ على جنوب(٩) اللان(١٠) ثمّ يمرّ على شمال بحر جرجان وبلاد خلج(١١) ثمّ يمرّ على جبل باب الأبواب(١٢) ثمّ يمرّ على وسط(١٣) بحر بنطس(١٤) ثمّ يمرّ

(١) هذا الإقليم: هذه الأقاليم [خ]. سقط من ع: هذا الإقليم

(٢) نحو من عشرة أجبال :من إحدى عشـ [ع]

(٣) سقط من ع: أربعين نهرا ومن

(٤) سقط من ع: جنوب

(٥) سقط من أ: ياجوج وماجوج

(٦) سقط من ع: سيستان

(٧) التغزغز: البربر [أ، ع، ف]، السعر [خ]، الثغرغز [ج، غ]، الثغر [ر]، بغرغر [ز، ك، ل]

(٨) كيماك: مكران [ف]

(٩) جنوب: جبال [ف]

(١٠) اللان: الان [ن]. سقط من خ: ثمّ يمرّ على جنوب اللان. سقط من ع: ثمّ بلاد كيماك ثمّ يمرّ على جنوب اللان

(١١) خلج: خليج [أ]، خلح [ج، خ، ط]، حانج [ف]، حلنج [ك]. ثمّ يمرّ على شمال بحر جرجان وبلاد خلج: يمرّ شمال بلاد كد [ع]

(١٢) وبلاد خلج ثمّ يمرّ على جبل باب الأبواب: ثمّ يمرّ على باب الأبواب وشمال بلاد الآن و [. . .] على شمال بلاد حليج فسطيط وشمال كشك وروس وبلغار [ع]

(١٣) وسط: جنوب [ل]

(١٤) بنطس: فلسطين [خ، ف]. وبلاد خلج ثمّ يمرّ على جبل باب الأبواب ثمّ يمرّ على وسط بحر بنطس: وشمال بلاد ماقدونية [ر]. زيادة في ز، ط، ف، ك، ل، ن: ثمّ يمرّ على جنوب بلاد برجان وشمال بلاد ماقدونية

على جنوب بحر الصقالبة[١] وجنوب جزيرة الوي[٢] وينتهي إلى بحر المغرب[٣] وأكثر أهل هذه البلدان ألوانهم الشقرة وهي كلّ مدينة عرضها من مز يه[٤] إلى مط[٥] أوّلها ممّا يلي المشرق.[٦]

أسماء المدن	الطول	العرض
بلاد ياجوج وماجوج		
بلاد اللان[٧] من الترك		
جزيرة في بحر المغرب		
جزيرة		
جزيرة		

(١) سقط من ز: جنوب بحر الصقالبة

(٢) الوي: الري [خ، ر]، البربر [غ]، اللون [ف]. سقط من ع: ثمّ يمرّ على وسط بحر بنطس ثمّ يمرّ على جنوب بحر الصقالبة وجنوب جزيرة الوي

(٣) وينتهي إلى بحر المغرب: وشمال بحر المغرب [غ]

(٤) مز يه: سبع وأربعين درجة وخمس عشرة دقيقة [ر]، مز درجة ويه دقيقة [غ]، مز درجة وخمسة عشرة دقيقة [ن]

(٥) مط: تسع واربعين درجة [ر]

(٦) سقط من ع: أكثر أهل هذه البلدان ألوانهم الشقرة وهي كلّ مدينة عرضها من مز يه إلى مط أوّلها ممّا يلي المشرق. زيادة في ك: والله أعلم. زيادة في ل: هذه اسماء المدن من الإقليم السابع

(٧) اللان: الطاق [أ]

فصل

واعلم يا أخي بأنّ في كلّ إقليم من هذه الأقاليم السبعة ألوف مدن تزيد وتنقص وفي كلّ مدينة أمم[١] من الناس مختلفة ألسنتهم وألوانهم[٢] وطباعهم وأخلاقهم وآرائهم[٣] ومذاهبهم وأعمالهم وصنائعهم[٤] وعاداتهم لا يشبه بعضهم بعضاً وهذا حكم حيوانها ومعادنها[٥] مختلفة الشكل والطعم واللون[٦] والرائحة وسبب ذلك اختلاف أهوية البلاد وتربة البقاع وعذوبة المياه وملوحتها وكلّ هذا الاختلاف بحسب طوالع البروج ودرجاتها على آفاق تلك البلاد وبحسب ممرّات الكواكب على مسامتات تلك البقاع ومطارحات[٧] شعاعاتها من الآفاق على (١٨٠) تلك المواضع[٨] وهذه جملة يطول شرحها وذكر أنّ ملكاً من[٩] الأوّلين أمر وقتاً من الزمان بأن يعدّ المدن[١٠] من الربع المسكون من الأرض فوجد سبعة عشر ألف مدينة وكسر سوى القرى.

---

(١) أمم: أمّة [ع]
(٢) سقط من ك: وألوانهم
(٣) سقط من ج: وآرائهم
(٤) زيادة في ط: وأخلاقهم
(٥) زيادة في ج: ونباتها
(٦) والطعم واللون: والطبع واللون والطعم [ن]
(٧) ومطارحات: ومطارح [خ، ر، ز، ط، ع، ف، ل،س، ن]
(٨) المواضع: البقاع والمواضع [ف]
(٩) زيادة في ع: الملوك في
(١٠) زيادة في ز، ك: المسكونة. سقط من خ: المدن

واعلم بأنّه[١] ربّما تزيد مدن الأرض[٢] وربّما[٣] ينقص[٤] عددها ويكون[٥] ذلك بحسب موجبات أحكام القرانات[٦] وأدوار الأفلاك[٧] وذلك أنّ القرانات[٨] الدالّة[٩] على قوّة السعود[١٠] واعتدال الزمان واستواء طبيعة الأركان ومجيء الأنبياء وتواتر الوحي وكثرة العلماء وعدل الملوك وصلاح أحوال الناس ونزول[١١] بركات السماء بالغيث فتزكو الأرض والنبات ويكثر تولّد الحيوان وتعمر البلاد ويكثر البنيان[١٢] والمدن[١٣] فالقرانات الدالّة على قوّة النحوس[١٤] وفساد الزمان وخروج المزاج[١٥] من الاعتدال وانقطاع الوحي وقلّة العلماء وموت الأخيار وجود الملوك[١٦] وفساد أخلاق الناس وسوء أعمالهم واختلاف آرائهم

(١) سقط من خ: بأنّه
(٢) زيادة في ف: وقتا
(٣) سقط من ز، ك: وربّما
(٤) زيادة في ف: وقتا
(٥) ويكون: وربّما يكون [ز، ك]
(٦) أحكام القرانات: الفلك أحكام النجوم والقرانات [ف]. موجبات أحكام القرانات: موجب القرانات [ل]
(٧) الأفلاك: الألوف [ج، خ، ط، ع، غ، ف]، الأفلاك والألوف [ك، ل]
(٨) سقط من ع: وأدوار الألوف وذلك أنّ القرانات
(٩) الدالّة: دالّة [ف]
(١٠) السعود: النحوس والسعود [ر]، السعادة [ل]، والسعد [ن]
(١١) ونزول: وحب نزول [ف]
(١٢) سقط من ع: ويكثر البنيان
(١٣) البنيان والمدن: بنيان المدن [ر، ز، ن، غ، ف، ك]. زيادة في ع، ف: والقرى
(١٤) قوّة النحوس: النحوس وقوّتها [ز]
(١٥) المزاج: الزمان [خ]
(١٦) الملوك: السلطان والملوك [ع]

ومنع نزول البركات من السماء بالغيث فلا تزكو[١] الأرض ويخفّ النبات ويهلك الحيوان وتخرب المدن في البلاد.

فصل[٢]

واعلم يا أخي بأنّ أمور هذه الدنيا دول ونوب تدور بين أهلها قرناً بعد قرن ومن أمّة إلى أمّة ومن بلد إلى بلد[٣]. واعلم بأنّ كلّ دولة لها وقت فيه يبتدئ وغاية إليها[٤] يرتقي[٥] وحدّ إليه ينتهي[٦] فإذا بلغت إلى أقصى غاياتها ومدى نهايتها[٧] أخذت في[٨] الانحطاط والنقصان وبدأ في أهلها الشؤم والخذلان واستأنف في الآخرين[٩] القوّة والنشاط والظهور[١٠] والانبساط وجعل كلّ يوم يقوى هذا ويزيد ويضعف هذا وينقص إلى أن يضمحلّ الأوّل

(١) تزكو: تزكوا [ج، خ، ط، ع، ف، ل]

(٢) سقط من ج، خ: فصل

(٣) ومن بلد إلى بلد: ومن أهل بلد إلى أهل بلد [ع]. زيادة في ع: ومن أهل بيت إلى أهل بيت. سقط من ر: واعلم يا أخي بأنّ أمور هذه الدنيا دول ونوب تدور بين أهلها قرنا بعد قرن ومن أمّة إلى أمّة ومن بلد إلى بلد. زيادة في ع: إلى أن ينتهي

(٤) سقط من ط: إليها

(٥) يرتقي: ينتهي [غ]

(٦) ينتهي: يرتقي [غ]

(٧) نهايتها: نهاياتها [خ، ز، ط]. غاياتها ومدى نهايتها: مدى غاياتها ونهايتها [ف]

(٨) أخذت في: تسارع إليه [ز، ك]

(٩) الآخرين: الآخرى [غ]

(١٠) سقط من ف: والظهور

المتقدّم ويستمكن[١] الثاني[٢] المتأخّر.

والمثال في ذلك مجاري أحكام الزمان وذلك أنّ[٣] الزمان[٤] نصفه نهار (١٨١) مضيء ونصفه ليل مظلم وأيضاً نصفه صيف[٥] حارّ ونصفه شتاء بارد وهما يتداولان في مجيئهما وذهابهما كلّما ذهب هذا رجع هذا وتارة يزيد هذا وتارة[٦] ينقص هذا وكلّما نقص من أحدهما زاد في الآخر بذلك المقدار حتّى إذا تناهيا إلى غاياتهما في الزيادة والنقصان[٧] ابتدأ النقص في الذي تناهى في الزيادة[٨] وابتدأت الزيادة[٩] في الذي تناهى[١٠] في النقصان ولا يزالان هكذا إلى أن يتساويا[١١] في مقدارهما ثمّ يتجاوزان على حالتيهما إلى أن تناهيا في غايتيهما[١٢] من الزيادة والنقصان وكلّما تناهى أحدهما في الزيادة ظهرت قوّته وكثرت أفعاله في العالم وخفي قوّة ضدّه وقلّت أفعاله.

---

(١) ويستمكن: وليستمكن [خ]، ويمكن [ر]، ويتمكّن [غ، ن]

(٢) الثاني: الحاوي [خ]، الآتي [ز، ك]، الجائي [غ، ف، ل، ن]

(٣) زيادة في خ: أحكام

(٤) زيادة في خ، ر، ف: كلّه. زيادة في ز، ك، ل: كلّه نصفان

(٥) صيف: ليل [ل]

(٦) سقط من ر، ز، غ، ط، ف، ك، ن: وتارة

(٧) زيادة في غ: سواء النقصان والزيادة

(٨) في الزيادة: في الطول والزيادة [ر]. سقط من غ: ابتدأ النقص في الذي تناهى في الزيادة

(٩) الزيادة: النقصان [خ]

(١٠) سقط من أ، ن: إلى غاياتهما في الزيادة والنقصان ابتدأ النقص في الذي تناهى في الزيادة وابتدأت الزيادة في الذي تناهى

(١١) أن يتساويا: أن ينتهي متساويا [ر]

(١٢) غايتيهما: حاليهما [ف]

فهذا حكم الزمان في دولة أهل الخير (أ ٣٩) ودولة أهل الشرّ[١] تارة تكون الدولة والقوّة وظهور الأفعال في العالم لأهل الخير[٢] وتارة تكون الدولة والقوّة وظهور الأفعال في العالم[٣] لأهل الشرّ[٤] كما ذكر الله عزّ وجلّ[٥] فقال[٦] ﴿وَتِلْكَ ٱلْأَيَّامُ نُدَاوِلُهَا بَيْنَ ٱلنَّاسِ﴾ ﴿وَمَا يَعْقِلُهَا إِلَّا العَالِمُونَ﴾.[٧]

فصل

فقد نرى[٨] أيّها الأخ البارّ الرحيم أيّدك الله وإيّانا بروح منه أنّه قد تناهت دولة أهل الشرّ وظهرت قوّتهم وكثرت أفعالهم في العالم بهذا الزمان وليس بعد التناهي في الزيادة إلّا الانحطاط والنقصان.[٩] واعلم بأنّ الدولة والملك ينتقلان في كلّ[١٠] دهر وزمان ودور وقران ومن أمّة إلى أمّة[١١] ومن أهل بيت إلى أهل

---

(١) سقط من خ: ودولة أهل الشرّ

(٢) سقط من خ، ط: تارة تكون الدولة والقوّة وظهور الأفعال في العالم لأهل الخير. الخير: الشرّ [ل]

(٣) سقط من ف: في العالم

(٤) الشرّ: الخير [ل]

(٥) عزّ وجلّ: تعالى [خ]، سبحانه وتعالى [ف]. ذكر الله عزّ وجلّ: قال الله تعالى [ل]

(٦) فقال: بقوله [غ]، وقال [ز، ك]. سقط من ف، ل: فقال

(٧) زيادة في ك: والله أعلم

(٨) نرى: ترى [خ، ر، ز، ن]

(٩) زيادة في ف: بهذا الزمان [ف]

(١٠) في كلّ: من [ن]

(١١) سقط من ف: إلى أمّة

بيت[١] ومن بلد إلى بلد. واعلم يا أخي بأنّ دولة[٢] أهل الخير[٣] يبدو أوّلها من قوم علماء[٤] حكماء أخيار فضلاء يجتمعون على رأي واحد ويتّفقون[٥] على دين واحد[٦] ويعقدون[٧] بينهم عهداً[٨] وميثاقاً ألّا يتجادلوا[٩] ولا يتقاعدوا[١٠] عن نصرة[١١] بعضهم بعضاً ويكونوا كرجل واحد[١٢] في جميع أمورهم وكنفس واحدة[١٣] في (١٨٢) جميع تدبيرهم فيما يقصدون من نصرة الدين وطلب[١٤] الآخرة لا يبتغون سوى وجه[١٥] الله جلّ ثناؤه ورضوانه ﴿جَزَاءً وَلَا شُكُوراً﴾.

فهل لك أيّها الأخ البارّ الرحيم أيّدك الله وإيّانا بروح منه بأن

---

(١) سقط من ج: إلى أهل بيت. من أهل بيت إلى أهل بيت: من بيت إلى بيت [ن]

(٢) سقط من ن: دولة

(٣) زيادة في ع: إنّما

(٤) سقط من غ: علماء

(٥) زيادة في ط: على مذهب واحد

(٦) على دين واحد: على مذهب واحد [ك]. زيادة في خ: ومذهب واحد كرجل واحد. زيادة في ر، ع، غ، ف، ل: ومذهب واحد

(٧) ويعقدون: ويعهدون [غ]

(٨) عهدا: عقدا [ع]

(٩) زيادة في ف: ولا يسافلون [ف]

(١٠) يتقاعدوا: يتجاوزون [ج]

(١١) نصرة: نصرة الدين [ل]

(١٢) كرجل واحد: كيد واحدة وكرجل واحدة [ع]

(١٣) سقط من خ: في جميع أمورهم وكنفس واحدة

(١٤) زيادة في ف: ثواب

(١٥) وجه: رحمة [ج]

ترغب في صحبة[١] إخوان لك نصحاء[٢] وأصدقاء لك[٣] أخيار فضلاء[٤] هذه صفتهم وأن تقصد مقصدهم[٥] وتتخلّق بأخلاقهم وتنظر في علومهم وتعرف مناهجهم وتكون معهم[٦] وتنجو ﴿بِمَفَازَاتِهِمْ[٧] لَا يَمَسُّهُمُ ٱلسُّوءُ وَلَا هُمْ[٨] يَحْزَنُونَ﴾.

وفّقك الله أيّها الأخ وجميع إخواننا للصواب بفضله ومنّه إنّه وليّ ذلك والقادر عليه.

تمّت رسالة جغرافيا وهي الرابعة والحمد لله حقّ حمده والصلوة على نبيّنا محمّد وآله وصحبه.

---

(١) صحبة: محبّة [خ]
(٢) نصحاء: فضلاء [غ]
(٣) زيادة في ف: كرام
(٤) فضلاء: نبلاء [غ]
(٥) مقصدهم: مقاصدهم [ك]. زيادة في ل: وتنحوا نحوهم
(٦) وتكون معهم: وتسلك سبيلهم [ع]. سقط من ف: معهم
(٧) بمفازاتهم: لمقاربتهم [ع]، بمفازتهم [ن]
(٨) سقط من ع: ولا هم

فهرس المصطلحات

فهرس الأماكن

- ت -

- ج -

- ح -

- خ -

- د -

- ز -

- س -

- ش -

- ص -

- ط -